WORKBOOK TO ACCOMPANY

# BEHAVIORAL STATISTICS IN ACTION



MARK VERNOY
Palomar College

JUDITH VERNOY

Wadsworth Publishing Company
Belmont, California
A Division of Wadsworth, Inc.

## CONTENTS






1 2 3 4 5 6 7 8 9 10—96 95 94 93 92

ISBN 0-534-16086-7

# CHAPTER 0

# SUCCEEDING IN (SURVIVING) STATISTICS

Our aim in writing a statistics textbook as well as this workbook is to introduce you to the uses of statistics in psychological research. Most psychology students do not understand why they are forced to take a *math* course when they are majoring in the study of behavior. Many enter their first statistics course with a great deal of anxiety and apprehension. They feel that at the very least, the course will be dreadfully boring, and at the worst, it will be extremely difficult. We have worked hard to write a textbook that will (hopefully!) allay these fears. After having attended your stat classes and having read your stat textbook, we hope you will leave this course not only with a sound knowledge of statistics but also with a positive attitude and an appreciation for the role of statistics in the field of psychology. This workbook introduction gives some sound advice that should aid you in your endeavor to learn and understand statistics. During the next few pages, we will discuss several things you can do in order to succeed in this class.

**BEFORE THE CLASS BEGINS**

*ASSESS YOUR MATHEMATICAL ABILITY*. Before you attend your first statistics class, you should do several things. First and foremost, you should assess your mathematical ability. Most psychological statistics classes list elementary or intermediate college algebra as a prerequisite. With a solid foundation in arithmetic and algebra, you should not find statistics difficult. To assess your mathematical ability, answer the following questions.

**1.** What symbol, when placed between 5 and 7, indicates that 5 is less than 7?

**2.** What does $Y \approx X$ mean?

**3.** What is the reciprocal of 2/6?

**4.** $|-23|$ means:

and it equals:

**5.** $x^2$ means:

**6.** $23 + (-3) =$

**7.** $20 - (-3) =$

**8.** $20 \cdot 2 =$

**9.** $20/2 =$

**10.** $20 \cdot (-2) =$

**11.** $20/(-2) =$

**12.** $(-10) \cdot (-3) =$

**13.** $(-10) \cdot (-4) \cdot (0) =$

**14.** $20 - 4(2/4) =$

**15.** $20 + (10 \cdot 6 - 2 + 13) =$

**16.** Does $xy^2 = (xy)^2$?

Answer yes or no, then explain your answer.

**17.** Does $(x + y)^2 = x^2 + y^2$?

Answer yes or no, then explain your answer.

**18.** $\sqrt{16x} =$

**19.** $\sqrt{49} =$

Given $X = 3$ and $Y = 2$, answer questions 20 through 25.

**20.** $2 \cdot X =$

**21.** $2X =$

**22.** $2X + 7 \cdot (4Y - 3) =$

**23.** $\frac{3X}{2Y} =$

**24.** $\frac{3X - 2Y}{\sqrt{5Y + 6}} =$

**25.** $\frac{3X}{2} \cdot \frac{4Y + 3}{25} =$

After working the preceding problems, check your answers. These answers can be found at the end of the workbook with the answers for the rest of the chapters. (If you have trouble with *ANY* of the preceding questions, you may have difficulty with this course. You will need to reevaluate your preparation for this course and discuss this with your professor as soon as possible.)

*BEGIN TO READ THE TEXTBOOK.* Before you attend the first lecture, you should scan the Table of Contents and read through Chapter 1 in the textbook. This will familiarize you with the topics that will possibly be covered in the course, and it may suggest some questions that you might ask during the first class session.

*TAKE A POSITIVE APPROACH TO THE COURSE.* Once you know your mathematical skills are up to speed and you have taken a brief look at the text, you need to convince yourself that you can succeed in this course. A positive, can-do, attitude will help you get past the rough spots in the course. But you must start with a strong positive attitude and continue that attitude throughout the course.

*SET PRIORITIES.* A positive approach alone will not carry you through this course. You must also set your studies as your highest priority. This means attending all classes (including review sessions, discussion groups, and labs) and setting aside a study time of at least two to four hours for each hour you spend in class. Promise yourself at the beginning of the class that you will work hard to succeed, and keep your promise. Remember that it is your responsiblity to learn the material. Don't make excuses or blame the book or your professor, and don't kid yourself about how well you study or how smart you are. If you study hard, attend all classes, and read the material before going to class so that you're prepared to ask questions about things you don't understand, you *will* learn statistics. So set your priorities, and work hard to succeed in statistics.

## ONCE THE CLASS BEGINS: IN CLASS

*ARRIVE AT THE FIRST CLASS EARLY AND SIT IN THE FRONT ROW.* In a statistics class, the best seat in the house is front row center. Since most students tend to "homestead" a particular chair or desk, it is advisable to claim the best seat at the first meeting of the class. So arrive early for the first few class meetings and homestead your front-row seat. You want to be at the front of the class so that you can hear the professor clearly and have an unobstructed view of the board. (If your professor uses an overhead projector, the best seats will be slightly to the left or right of the projector, not directly in back of it.) Sitting in the front row also forces you to become more involved in the class. You will share more eye contact with the professor, and your questions will be answered more quickly than those in the rear of the lecture hall.

*ATTEND ALL CLASSES.* Statistics is a class that is sequential. Almost every day you are taught a new skill or formula that is necessary for the understanding of the skill or formula that will be taught the next day. Because of the sequential nature of the course, it is imperative that you attend each and every class. Get sick or go to the beach between semesters, but attend all statistics lectures. If for some reason you do miss a lecture, make sure you find out what was covered and you understand what was covered. If you know you will miss the class in advance, you can ask a friend to tape record the lecture, with the professor's permission. Never ask your professor if he or she is going to cover anything important. Professors feel that *everything* they cover is important. Assume that you are going to miss something critical, and plan accordingly. After you have studied a friend's notes and have possibly listened to the lecture on tape, write down any questions you have and discuss these with another class member or the professor. If you get behind, it is your responsibility to catch up immediately.

*TAKE DETAILED CLASS NOTES.* The strongest memory aids that you will have in this course are the text and your class notes. Take detailed class notes. Try to follow everything the professor does and says, and try to get this information down on paper. If the professor works a problem on the board, copy it down for later reference. (You might even want to tape record the lectures, with the professor's permission.) Also, make sure that your notes are clean and readable. Notes with lots of arrows and scratchouts should be recopied as soon as possible.

*WORK ALL IN-CLASS PROBLEMS ALONG WITH THE PROFESSOR.* It is impossible, as far as we know, to teach statistics without working statistical problems in class. An effective way to practice statistics is to work the demonstration problems used by your professor in class. Don't just copy down his/her example; compute the numbers yourself. If you get a different number from your professor, ask a question at once to reconcile the difference. (Remember, professors are human--at least, most of them are--and they make mistakes too.)

*ASK QUESTIONS.* There is nothing worse than being in a statistics class and not understanding a word that the professor is saying. The professor is there to teach you statistics, but it is your responsibility as a student to learn statistics. At the beginning of each class period, it is quite legitimate to ask a question about the topics discussed in the previous class session. If the professor is covering a topic too quickly or if you do not understand a concept, ask about it. Remember, it is *your* responsibility to learn the material.

## ONCE THE CLASS BEGINS: OUTSIDE OF CLASS

*READ THE BOOK BEFORE THE MATERIAL IS COVERED IN CLASS.* Your professor will give you a list of chapters or a list of topics that will be covered in his/her lectures at the beginning of the course. Use this list to pace your reading so that you are always ahead of the class lecture--read the chapter in the book before it is discussed in class. When you read each chapter, have a pencil and calculator handy so that you can take notes and work the problems along with the text. This will give you some practice in solving the types of problems your professor will discuss, introduce you to the concepts that will be covered, and allow you to take more effective class notes.

*REVIEW THE BOOK AFTER THE MATERIAL IS COVERED IN CLASS.* After your professor has covered the material discussed in the book, it is a good idea to briefly review the book chapter to reinforce the lecture. This will also give you a chance to reconcile any differences in notation between the the text and your professor.

*REVIEW YOUR CLASS NOTES.* After each class session, it is important to review your notes. This review will give you the necessary preparation for the assigned homework as well as allow you to recopy or clean up the notes for future reference.

*PRACTICE WORKING STATISTICAL PROBLEMS.* Once you have reviewed the latest lecture and the relevant chapter in the book, you will need to practice using the statistics you have just learned. The way to practice statistics is to work problems. Professors assign homework problems to give you practice with statistics they have recently covered; but assigned homework is the *minimum* amount of practice you should be getting. To be certain that you really understand statistical procedures and concepts, it is best to do more practice than you get from merely doing the homework. There are several sources of practice problems other than assigned homework. The textbook and this workbook have problems that are designed to give you practice on all statistical procedures covered in this course. The text and this workbook have a total of at least 40, and as many as 80, different problems for each chapter. These problems, plus any additional problems you might make up yourself, should give you sufficient practice.

*GET HELP QUICKLY.* If you have difficulty understanding the text or if you have trouble doing your homework, don't just sit there for hours stewing over the difficulty. GET HELP! If you can't figure out a problem or a particular statistical concept within half an hour or so, you certainly won't figure it out in three or four hours. There are many sources for help: the textbook, your class notes, taped lectures, friends in the class, friends who have previously taken the class, tutors, teaching and lab assistants, and the professor. Try to cultivate as many sources of competent help as you can. Make friends in class immediately. Talk to the people who sit near you in class and exchange telephone numbers. Find out during the first class period if there is tutorial help. Get the names and telephone numbers of any lab or teaching assistants. Ask for the professor's office hours and his/her campus telephone extension. Never let a problem reduce you to a frustrated bundle of nerves; there's an abundance of people who are willing and able to clarify things for you.

## TESTS

*PREPARING FOR A TEST.* The best way to prepare for a statistics test is to attend every class session, take good class notes, ask questions in class, read the text, work all homework problems, work extra problems from the text or the workbook, and get help when you need it. If you keep up with the class, you should be prepared to take a test at any time. The only additional thing you will need to do the days before the test is to work additional problems to make sure you understand all the statistical concepts, procedures, and formulas. If your professor allows open-book or open-note tests, you should take the time to organize all the important concepts, formulas, and procedures on one or two pages for easy access during the test. It is also to your advantage to quiz your professor about the test: Is the test open-book or open-note? What topics will be covered? How many questions will there be? How much time will be allowed? How many questions will be devoted to a particular topic? What exactly are the questions, and what are the answers to the questions? (Don't expect your professor to give you the questions and answers ahead of time, but it can't hurt to ask.) Test anxiety and "blanking out" during a test are both the result of poor preparation and low self-confidence. If a concept or procedure is not clear to you, the stress of taking a test will make it even fuzzier. You must begin preparing for the first test from the very first day of class. And finally, get a good night's sleep before the test. Everyone performs intellectual tasks better when they are rested.

*TAKING A TEST.* Taking a test is stressful for anyone. Properly preparing for a test helps, but there are some other things you can do to further reduce the stress of the test. Arrive early so you can be sure to get the seat you have "homesteaded" throughout the semester. Make sure you have all the necessary materials--pencils, paper, a calculator with fresh batteries, and any notes that are legal to use and that you feel might be helpful. Once you have the test in hand, take a few seconds to read the entire test so that you can estimate how long each question will take and you can determine which questions might be the most difficult for you. Also look to see if information from one question will be needed to answer questions that follow. As you begin to solve the problems, it is important to write out all formulas and intermediate results as time permits. If you are taking an open-book or open-note test, check all the formulas. Always use your calculator--try to avoid doing any calculations in your head. Work as quickly and as carefully as possible. If you are stumped by a problem, don't waste time on it. Leave it for the end. When you finish the test, your first impulse may be to escape. But if you finish the test before the time limit has expired, use the extra time to go back and check or rework each problem, beginning with the most difficult problems. Remember that no professor has, to our knowledge, ever given extra credit to students who finish before the time limit has expired. You have the right to check your work for errors.

*WHEN YOUR PROFESSOR RETURNS YOUR TEST.* The first thing you should do when you get your test back from your professor is to make sure that the grade is accurate. If you have received points for each problem, add them all up to verify that there were no errors in addition. Next, you should take a close look at your incorrect problems to make sure you understand your mistakes. If you can't figure out the problem by yourself, get help. Make sure that you rework all the missed problems so you will have a perfect test to study from when you take the final. You will want to discuss the test with your professor if you think there was some mistake in grading. Finally, you should use the test as a tool to evaluate your study techniques. The test will help you identify those areas for which you were poorly prepared. Once those problem areas are identified, you will want to modify your study habits to correct those deficiencies.

**A FINAL NOTE**

No one can guarantee you a good grade in statistics, but if you are willing to put in the time to learn the material, you should not only enjoy but succeed in the course. The aforementioned hints can help you learn the material, but it is up to you to make learning statistics your number one priority. Maintain a positive attitude toward the course throughout the term, stay on top of the material, work lots of problems, and get help when you need it.

# CHAPTER 1

# AN INTRODUCTION TO STATISTICS

Round off the numbers in problems 1 through 5 to three digits to the right of the decimal point.

1. 234.25467
2. 1.766666
3. 8.8976
4. 4567.98734
5. 777.98421
6. 29.888888
7. 99.9999
8. 1234.5678
9. 987.9544
10. 0.0004

Generate a research hypothesis and a null hypothesis for the research questions stated in problems 11 through 16.

**11.** Is there a difference between the number of personality traits exhibited by 3-year-olds as compared to adults?

Research hypothesis:

Null hypothesis:

**12.** Can human performance be enhanced using mind-altering drugs?

Research hypothesis:

Null hypothesis:

**13.** Are the women of one culture more aggressive than the women of another culture?

Research hypothesis:

Null hypothesis:

**14.** Does stress affect the level of a specific nervous system hormone?

Research hypothesis:

Null hypothesis:

**15.** Do people behave differently when they know that they are being observed as compared to when they believe they are not being observed?

Research hypothesis:

Null hypothesis:

**16.** Is one method of teaching preschoolers better than another method?

Research hypothesis:

Null hypothesis:

**Identify the independent and dependent variables in the experiments described in problems 17 through 21.**

**17.** A cognitive psychologist wishes to see if subjects can remember more items from a list of words as compared to a list of numbers.

Independent variable:

Dependent variable:

**18.** A social psychologist is interested in measuring the differences in social interaction (conversation) in small versus large groups.

Independent variable:

Dependent variable:

**19.** A health psychologist asks one group of patients to use one type of relaxation technique and another group to use a different type of relaxation technique to see if they can lower their blood pressure.

Independent variable:

Dependent variable:

**20.** Two groups of monkeys are given injections of different levels of a particular hormone to see if the hormone has an effect on the mating behavior of the monkeys.

Independent variable:

Dependent variable:

**21.** A psychologist is interested in seeing if a certain type of mathematical problem can be solved faster using a new problem-solving technique as compared with an existing, standard technique.

Independent variable:

Dependent variable:

Tell whether the examples in problems 22 through 31 represent the nominal, ordinal, interval, or ratio scale of measurement.

**22.** Socioeconomic status (lower, middle, upper)

Scale:

**23.** The number of errors a rat makes running a maze

Scale:

**24.** Temperature in degrees centigrade

Scale:

**25.** The number of firings per minute from a single neuron

Scale:

**26.** The perceived distance of an object

Scale:

**27.** The current world rankings of chess masters

Scale:

**28.** The amount of salivation, in milliliters, of Pavlov's dog

Scale:

**29.** The time it takes to solve a puzzle

Scale:

**30.** The average number of words used in a sentence by a specific writer

Scale:

**31.** The number of confederates in a group

Scale:

CHAPTER 2

# FREQUENCY DISTRIBUTIONS

Create rank distributions for the data shown in problems 1 through 5.

1. Reaction times in milliseconds:

   123 234 234 567 236 129 438 283 444
   358

2. Achievement test scores:

   78 36 67 23 98 56 24 76 92 43 22
   10 25 67

3. Number of targets hit out of 100 in a video game:

29 98 67 32 56 98 66 30 42 44 23
10 10

4. Time spent daydreaming each day in minutes:

123 80 23 10 65 236 34 67 98 45 22
15 67 8

5. Number of aggressive acts per hour:

3 5 4 6 9 3 5 2 9 8 10 23 13 5
4 6 7 9

Problems 6 through 10 present a range of scores and the desired number of intervals. You are to compute the appropriate interval size to generate that number of intervals.

**6.** The number of aggressive acts per hour at the local college preschool ranges from a low of 1 to a high of 30. The desired number of intervals is 10.

Interval Size =

**7.** The scores on a personality test range from 39 to 173. The desired number of intervals is 15.

Interval Size =

**8.** The time it takes subjects to say the name of a color ranges from 600 to 1139 milliseconds. The desired number of intervals is 12.

Interval Size =

**9.** The number of trials it takes to classically condition a dog to salivate to a tone ranges from 56 to 120 trials. The desired number of intervals is 13.

Interval Size =

**10.** The number of words spoken by a 2-year-old ranges from 50 to 800. The desired number of intervals is 15.

Interval Size =

**A physiological psychologist has found that the level of a particular hormone in the blood of humans ranges from 750 micrograms per liter of blood to 1500 micrograms per liter. Use this information to indicate the approximate number of intervals in her frequency distribution when the interval size is equal to the numbers depicted in problems 11 through 15. Also indicate by circling *Yes* or *No* whether this number of intervals is acceptable.**

**11.** 80 micrograms

**Number of Intervals =**

**Acceptable? Yes No**

**12.** 75 micrograms

**Number of Intervals =**

**Acceptable? Yes No**

**13.** 35 micrograms

**Number of Intervals =**

**Acceptable? Yes No**

**14.** 50 micrograms

**Number of Intervals =**

**Acceptable? Yes No**

**15.** 45 micrograms

**Number of Intervals =**

**Acceptable? Yes No**

**For the data in problems 16 through 25, use the given interval size to generate a frequency distribution that includes the real limits, the apparent limits, the frequency, the relative frequency, the cumulative frequency, and the cumulative relative frequency.**

**16.** **The following scores are reaction times in milliseconds. Use an interval size of 40 milliseconds.**

| | | | | | | | | |
|---|---|---|---|---|---|---|---|---|
| 123 | 234 | 234 | 567 | 236 | 129 | 438 | 283 | 444 |
| 358 | 445 | 125 | 435 | 566 | 542 | 234 | 123 | 459 |
| 365 | 512 | 321 | 457 | 338 | 198 | 319 | | |

| *Real Lim* | *Appl Lim* | *F* | *RF* | *CF* | *CRF* |
|---|---|---|---|---|---|

**17.** The following scores are achievement test scores. Use an interval size of 5 points.

| | | | | | | | | | | |
|---|---|---|---|---|---|---|---|---|---|---|
| 99 | 98 | 98 | 97 | 95 | 93 | 92 | 92 | 91 | 90 | 90 |
| 89 | 88 | 88 | 86 | 85 | 83 | 83 | 80 | 79 | 79 | 79 |
| 76 | 75 | 74 | 73 | 71 | 71 | 71 | 71 | 70 | 70 | 70 |
| 68 | 67 | 65 | 65 | 65 | 65 | 65 | 64 | 63 | 60 | 58 |
| 58 | 56 | 55 | 54 | 45 | 40 | | | | | |

| *Real Lim* | *Appl Lim* | *F* | *RF* | *CF* | *CRF* |
|---|---|---|---|---|---|

**18.** The following scores are the number of targets hit out of 100 by 40 elementary school children playing a video game. Use an interval size of 6 hits.

| | | | | | | | | | | |
|---|---|---|---|---|---|---|---|---|---|---|
| 12 | 14 | 17 | 18 | 19 | 23 | 25 | 28 | 34 | 35 | 39 |
| 40 | 42 | 47 | 48 | 50 | 53 | 55 | 56 | 59 | 60 | 65 |
| 69 | 71 | 73 | 76 | 78 | 78 | 80 | 82 | 83 | 83 | 83 |
| 84 | 85 | 85 | 86 | 89 | 90 | 98 | | | | |

| *Real Lim* | *Appl Lim* | *F* | *RF* | *CF* | *CRF* |
|---|---|---|---|---|---|

**19.** Thirty-five subjects reported the number of minutes they spent daydreaming during the past day. Use an interval size of 7 minutes.

| | | | | | | | | | | |
|---|---|---|---|---|---|---|---|---|---|---|
| 12 | 14 | 16 | 18 | 19 | 19 | 20 | 22 | 23 | 24 | 35 |
| 36 | 39 | 42 | 47 | 49 | 49 | 50 | 51 | 52 | 56 | 59 |
| 60 | 66 | 68 | 69 | 72 | 73 | 75 | 78 | 80 | 82 | 84 |
| 87 | 88 | | | | | | | | | |

| *Real Lim* | *Appl Lim* | *F* | *RF* | *CF* | *CRF* |
|---|---|---|---|---|---|

20. The number of aggressive acts per hour for 60 children is shown below. Use an interval size of 2 acts.

| | | | | | | | | | | |
|---|---|---|---|---|---|---|---|---|---|---|
| 1 | 2 | 2 | 2 | 2 | 3 | 4 | 4 | 5 | 5 | 6 |
| 7 | 7 | 7 | 7 | 8 | 8 | 8 | 9 | 10 | 10 | 10 |
| 11 | 11 | 12 | 12 | 12 | 12 | 12 | 13 | 13 | 14 | 15 |
| 15 | 15 | 15 | 16 | 17 | 18 | 18 | 18 | 18 | 19 | 19 |
| 19 | 19 | 19 | 20 | 20 | 20 | 20 | 21 | 22 | 23 | 23 |
| 23 | 23 | 24 | 24 | 26 | | | | | | |

| *Real Lim* | *Appl Lim* | *F* | *RF* | *CF* | *CRF* |
|---|---|---|---|---|---|

**21.** The number of positive statements about their self-image was recorded for 25 patients. Use an interval size of 2 statements.

2 2 3 4 4 4 4 4 4 4 5 5 5 5 5
5 6 6 6 7 8 9 10 18 26

| *Real Lim* | *Appl Lim* | *F* | *RF* | *CF* | *CRF* |
|---|---|---|---|---|---|

**22.** Ratings follow for 33 married people who rated their spouses on degree of attractiveness. Use an interval size of 35 points.

| | | | | | | | | |
|---|---|---|---|---|---|---|---|---|
| 500 | 500 | 523 | 524 | 550 | 600 | 625 | 629 | 630 |
| 650 | 675 | 690 | 700 | 750 | 780 | 799 | 800 | 837 |
| 840 | 850 | 850 | 850 | 860 | 870 | 880 | 890 | 900 |
| 900 | 910 | 920 | 930 | 940 | 950 | | | |

| *Real Lim* | *Appl Lim* | *F* | *RF* | *CF* | *CRF* |
|---|---|---|---|---|---|

**23.** Following is a list of the life change units (a score reflecting the amount of stress experienced as a result of stressful life events) reported by 50 subjects. Use an interval size of 25 life change units.

| | | | | | | | | |
|---|---|---|---|---|---|---|---|---|
| 150 | 175 | 375 | 210 | 400 | 216 | 300 | 175 | 374 |
| 163 | 263 | 152 | 176 | 185 | 192 | 197 | 233 | 216 |
| 241 | 221 | 232 | 316 | 233 | 357 | 321 | 368 | 300 |
| 277 | 298 | 254 | 274 | 216 | 285 | 219 | 276 | 222 |
| 245 | 264 | 233 | 361 | 242 | 304 | 251 | 176 | 221 |
| 165 | 212 | 222 | 196 | 196 | | | | |

| *Real Lim* | *Appl Lim* | *F* | *RF* | *CF* | *CRF* |
|---|---|---|---|---|---|

**24.** Following is the number of affirmative answers on a personality inventory. Use an interval size of 5 answers.

| | | | | | | | | | | |
|---|---|---|---|---|---|---|---|---|---|---|
| 20 | 22 | 25 | 25 | 25 | 27 | 28 | 31 | 33 | 35 | 37 |
| 37 | 38 | 42 | 43 | 44 | 45 | 46 | 48 | 49 | 50 | 51 |
| 55 | 56 | 58 | 59 | 61 | 63 | 64 | 67 | 69 | 70 | 71 |
| 72 | 74 | 76 | 78 | 79 | 82 | 84 | 84 | 86 | 87 | 89 |

| *Real Lim* | *Appl Lim* | *F* | *RF* | *CF* | *CRF* |
|---|---|---|---|---|---|

**25.** Following is a list of the number of uses reported by each subject when asked to name as many uses for a brick as possible. Use an interval size of 3 uses.

| | | | | | | | | | | |
|---|---|---|---|---|---|---|---|---|---|---|
| 5 | 5 | 6 | 7 | 8 | 8 | 9 | 9 | 9 | 9 | 10 |
| 10 | 10 | 10 | 10 | 13 | 13 | 14 | 15 | 15 | 16 | 18 |
| 19 | 19 | 19 | 21 | 23 | 24 | 22 | 26 | 28 | 29 | 30 |
| 31 | 32 | 33 | 34 | 35 | 38 | 48 | | | | |

| *Real Lim* | *Appl Lim* | *F* | *RF* | *CF* | *CRF* |
|---|---|---|---|---|---|

# CHAPTER 3

## GRAPHS

1. When drawing a graph of information portrayed in a frequency distribution, the frequency is usually represented along the __________ axis and the dependent variable measure is usually represented along the __________ axis.

2. When plotting a frequency polygon, the frequency is plotted at the __________ of the interval.

3. When plotting a frequency histogram, the sides of each bar are plotted at the __________ of the interval.

4. The points in a cumulative frequency polygon or a cumulative relative frequency polygon are always plotted at the __________ of the interval.

5. Use the following statistics aptitude test scores for 100 students to generate a frequency distribution. Use an interval size of 3 points.

| | | | | | | | | | | |
|---|---|---|---|---|---|---|---|---|---|---|
| 55 | 55 | 56 | 57 | 58 | 59 | 59 | 59 | 60 | 60 | 61 |
| 61 | 61 | 61 | 62 | 63 | 63 | 63 | 63 | 63 | 63 | 65 |
| 66 | 66 | 67 | 67 | 67 | 67 | 68 | 69 | 70 | 70 | 71 |
| 71 | 71 | 71 | 71 | 71 | 73 | 73 | 73 | 73 | 73 | 77 |
| 77 | 77 | 77 | 78 | 78 | 78 | 79 | 79 | 79 | 79 | 80 |
| 81 | 82 | 83 | 83 | 83 | 83 | 83 | 83 | 83 | 84 | 85 |
| 87 | 88 | 88 | 88 | 88 | 89 | 90 | 90 | 90 | 90 | 91 |
| 91 | 91 | 91 | 92 | 93 | 94 | 94 | 94 | 95 | 95 | 95 |
| 96 | 96 | 96 | 96 | 97 | 97 | 97 | 97 | 98 | 98 | 99 |
| 99 | | | | | | | | | | |

| *App Lim* | *F* | *CF* | *RF* | *CRF* |
|---|---|---|---|---|

Use the frequency distribution you created in problem number 5 to answer problems 6 through 10.

6. In the space below, plot a frequency histogram.

7. In the space below, plot a frequency polygon.

8. In the space below, plot a cumulative frequency polygon.

9. In the space below, plot a relative frequency polygon.

**10.** In the space below, plot a cumulative relative frequency polygon.

**Use the following frequency distribution to complete problems 11 through 15.**

| App Lim | F | CF | RF | CRF |
|---|---|---|---|---|
| 240 - 254 | 1 | 200 | .005 | 1.000 |
| 225 - 239 | 10 | 199 | .050 | .995 |
| 210 - 224 | 14 | 189 | .070 | .945 |
| 195 - 209 | 27 | 175 | .135 | .875 |
| 180 - 194 | 29 | 148 | .145 | .740 |
| 165 - 179 | 56 | 119 | .280 | .595 |
| 150 - 164 | 29 | 63 | .145 | .315 |
| 135 - 149 | 22 | 34 | .110 | .170 |
| 120 - 134 | 9 | 12 | .045 | .060 |
| 105 - 119 | 3 | 3 | .015 | .015 |

**11.** In the space below, plot a cumulative frequency polygon.

**12.** In the space below, plot a relative frequency polygon.

**13.** In the space below, plot a frequency histogram.

**14.** In the space below, plot a frequency polygon.

**15.** In the space below, plot a cumulative relative frequency polygon.

Use the following frequency distribution to complete problems 16 through 20.

| *App Lim* | *F* | *CF* | *RF* | *CRF* |
|---|---|---|---|---|
| 75 - 79 | 1 | 100 | .01 | 1.00 |
| 70 - 74 | 5 | 99 | .05 | .99 |
| 65 - 69 | 7 | 94 | .07 | .94 |
| 60 - 64 | 14 | 87 | .14 | .87 |
| 55 - 59 | 15 | 73 | .15 | .73 |
| 50 - 54 | 28 | 58 | .28 | .58 |
| 45 - 49 | 14 | 30 | .14 | .30 |
| 40 - 44 | 11 | 16 | .11 | .16 |
| 35 - 39 | 4 | 5 | .04 | .05 |
| 30 - 34 | 1 | 1 | .01 | .01 |

**16.** In the space below, plot a cumulative relative frequency polygon.

17. In the space below, plot a cumulative frequency polygon.

18. In the space below, plot a relative frequency polygon.

**19.** In the space below, plot a frequency polygon.

**20.** In the space below, plot a frequency histogram.

Use the following frequency distribution to complete problems 21 through 25.

| *App Lim* | *F* | *CF* | *RF* | *CRF* |
|---|---|---|---|---|
| 77 - 83 | 5 | 500 | .01 | 1.00 |
| 70 - 76 | 25 | 495 | .05 | .99 |
| 63 - 69 | 35 | 470 | .07 | .94 |
| 56 - 62 | 45 | 435 | .09 | .87 |
| 49 - 55 | 55 | 390 | .11 | .78 |
| 42 - 48 | 110 | 335 | .22 | .67 |
| 35 - 41 | 100 | 225 | .20 | .45 |
| 28 - 34 | 60 | 125 | .12 | .25 |
| 21 - 27 | 40 | 65 | .08 | .13 |
| 14 - 20 | 20 | 25 | .04 | .05 |
| 7 - 13 | 5 | 5 | .01 | .01 |

**21.** In the space below, plot a frequency polygon.

22. In the space below, plot a cumulative frequency polygon.

23. In the space below, plot a frequency histogram.

24. In the space below, plot a cumulative relative frequency polygon.

25. In the space below, plot a relative frequency polygon.

CHAPTER 4

# MEASURES OF CENTRAL TENDENCY

Use the following achievement test scores to answer problems 1 through 4.

| | | | | | | | | | | |
|---|---|---|---|---|---|---|---|---|---|---|
| 98 | 98 | 98 | 97 | 95 | 93 | 92 | 92 | 91 | 90 | 90 |
| 89 | 88 | 88 | 86 | 85 | 83 | 83 | 80 | 80 | 79 | 79 |

**1.** Compute the mean.

Mean =

**2.** Find the median.

Median =

**3.** Find the mode.

Mode =

**4.** Given the mean, the median, and the mode, what is the skew of this distribution?

Skew =

Use the following video game scores for 20 elementary school children to answer problems 5 through 8.

12 14 14 18 19 23 25 28 34 35 39
40 42 47 48 50 53 59 59 59

**5.** Compute the mean.

Mean =

**6.** Find the median.

Median =

**7.** Find the mode.

Mode =

**8.** Given the mean, the median, and the mode, what is the skew of this distribution?

Skew =

The following data represent the number of minutes each day that 15 subjects reported daydreaming. Please use these data to answer problems 9 through 12.

12 14 16 18 19 19 20 22 23 24 35
36 39 42 47

**9.** Compute the mean.

Mean =

**10.** Find the median.

Median =

**11.** Find the mode.

Mode =

**12.** Given the mean, the median, and the mode, what is the skew of this distribution?

Skew =

> Use the following data consisting of the number of aggressive acts per hour for 33 preschool children to answer problems 13 through 16.

| | | | | | | | | | | |
|---|---|---|---|---|---|---|---|---|---|---|
| 1 | 2 | 2 | 2 | 2 | 3 | 4 | 4 | 5 | 5 | 6 |
| 7 | 7 | 7 | 7 | 8 | 8 | 8 | 9 | 10 | 10 | 10 |
| 11 | 11 | 12 | 12 | 12 | 12 | 12 | 13 | 13 | 14 | 15 |

**13.** Compute the mean.

Mean =

**14.** Find the median.

Median =

**15.** Find the mode.

Mode =

**16.** Given the mean, the median, and the mode, what is the skew of this distribution?

Skew =

| The number of positive statements about their self-image was recorded for 25 patients. Use these data to answer problems 17 through 20. |
|---|

| | | | | | | | | | | |
|---|---|---|---|---|---|---|---|---|---|---|
| 2 | 2 | 3 | 4 | 4 | 4 | 4 | 4 | 4 | 4 | 5 |
| 5 | 5 | 5 | 5 | 5 | 6 | 6 | 6 | 7 | 8 | 9 |
| 10 | 18 | 26 | | | | | | | | |

**17.** Compute the mean.

Mean =

**18.** Find the median.

Median =

**19.** Find the mode.

Mode =

**20.** Given the mean, the median, and the mode, what is the skew of this distribution?

Skew =

Use the frequency distribution below to complete problems 21 through 24.

| *App Lim* | *F* | *CF* | *X* | *F•X* |
|---|---|---|---|---|
| 240 - 254 | 1 | 200 | | |
| 225 - 239 | 10 | 199 | | |
| 210 - 224 | 14 | 189 | | |
| 195 - 209 | 27 | 175 | | |
| 180 - 194 | 29 | 148 | | |
| 165 - 179 | 56 | 119 | | |
| 150 - 164 | 29 | 63 | | |
| 135 - 149 | 22 | 34 | | |
| 120 - 134 | 9 | 12 | | |
| 105 - 119 | 3 | 3 | | |

$\Sigma(F \cdot X) =$

**21.** Compute the mean.

Mean =

**22.** Compute the median.

Median =

**23.** Find the mode.

Mode =

**24.** Given the mean, the median, and the mode, what is the skew of this distribution?

Skew =

Use the frequency distribution below to answer questions 25 through 28.

| *App Lim* | *F* | *CF* | *X* | *F•X* |
|---|---|---|---|---|
| 75 - 79 | 1 | 100 | | |
| 70 - 74 | 5 | 99 | | |
| 65 - 69 | 7 | 94 | | |
| 60 - 64 | 14 | 87 | | |
| 55 - 59 | 15 | 73 | | |
| 50 - 54 | 28 | 58 | | |
| 45 - 49 | 14 | 30 | | |
| 40 - 44 | 11 | 16 | | |
| 35 - 39 | 4 | 5 | | |
| 30 - 34 | 1 | 1 | | |

$\Sigma(F \cdot X) =$

**25.** Compute the mean.

Mean =

**26.** Compute the median.

Median =

**27.** Find the mode.

Mode =

**28.** Given the mean, the median, and the mode, what is the skew of this distribution?

Skew =

CHAPTER 5

# MEASURES OF VARIABILITY

Use the following reaction time scores to answer problems 1 through 3.

| $X$ | $X^2$ | $X-\overline{X}$ | $(X-\overline{X})^2$ |
|---|---|---|---|
| 223 | | | |
| 234 | | | |
| 234 | | | |
| 567 | | | |
| 236 | | | |
| 129 | | | |
| 438 | | | |
| 283 | | | |
| 444 | | | |
| 358 | | | |
| 445 | | | |

$\Sigma X=$ $\Sigma X^2=$ $\Sigma(X-\overline{X})^2=$

**1.** Compute the mean.

Mean =

2. Compute the variance.

   Variance =

3. Compute the standard deviation.

   Standard deviation =

Use the following achievement test scores to answer problems 4 through 6.

| $X$ | $X^2$ | $X-\overline{X}$ | $(X-\overline{X})^2$ |
|---|---|---|---|
| 99 | | | |
| 98 | | | |
| 98 | | | |
| 97 | | | |
| 95 | | | |
| 93 | | | |
| 92 | | | |
| 92 | | | |
| 91 | | | |
| 90 | | | |
| 90 | | | |
| 89 | | | |
| 88 | | | |
| 88 | | | |
| 86 | | | |
| 85 | | | |
| 83 | | | |

$\Sigma X=$ $\Sigma X^2=$ $\Sigma(X-\overline{X})^2=$

**4.** Compute the mean.

Mean =

**5.** Compute the variance.

Variance =

**6.** Compute the standard deviation.

Standard deviation =

Use the following video game scores for 10 elementary school children to answer problems 7 through 9.

| $X$ | $X^2$ | $X-\overline{X}$ | $(X-\overline{X})^2$ |
|---|---|---|---|
| 12 | | | |
| 14 | | | |
| 17 | | | |
| 18 | | | |
| 19 | | | |
| 23 | | | |
| 25 | | | |
| 28 | | | |
| 34 | | | |
| 59 | | | |

$\Sigma X=$ $\Sigma X^2 =$ $\Sigma(X-\overline{X})^2 =$

**7.** Compute the mean.

Mean =

**8.** Compute the variance.

Variance =

**9.** Compute the standard deviation.

Standard deviation =

> **The following data represent the number of minutes spent daydreaming each day, as reported by 15 subjects. Please use these data to answer problems 10 through 12.**

| $X$ | $X^2$ | $X-\overline{X}$ | $(X-\overline{X})^2$ |
|---|---|---|---|
| 12 | | | |
| 14 | | | |
| 16 | | | |
| 18 | | | |
| 19 | | | |
| 19 | | | |
| 20 | | | |
| 22 | | | |
| 23 | | | |
| 24 | | | |
| 35 | | | |
| 36 | | | |
| 39 | | | |
| 42 | | | |
| 47 | | | |

$\Sigma X=$ $\Sigma X^2 =$ $\Sigma(X-\overline{X})^2 =$

**10.** Compute the mean.

Mean =

**11.** Compute the variance.

Variance =

**12.** Compute the standard deviation.

Standard deviation =

Use the following data on the number of aggressive acts per hour for 13 preschool children to answer problems 13 through 15.

| $X$ | $X^2$ | $X-\overline{X}$ | $(X-\overline{X})^2$ |
|---|---|---|---|
| 1 | | | |
| 2 | | | |
| 2 | | | |
| 2 | | | |
| 2 | | | |
| 7 | | | |
| 7 | | | |
| 7 | | | |
| 8 | | | |
| 9 | | | |
| 10 | | | |
| 12 | | | |
| 15 | | | |

$\Sigma X=$ $\Sigma X^2 =$ $\Sigma(X-\overline{X})^2 =$

**13.** Compute the mean.

Mean =

**14.** Compute the variance.

Variance =

**15.** Compute the standard deviation.

Standard deviation =

The number of positive statements about their self-image was recorded for 15 patients. Use these data to answer problems 16 through 18.

| $X$ | $X^2$ | $X-\overline{X}$ | $(X-\overline{X})^2$ |
|---|---|---|---|
| 2 | | | |
| 2 | | | |
| 3 | | | |
| 4 | | | |
| 5 | | | |
| 5 | | | |
| 6 | | | |
| 7 | | | |
| 8 | | | |
| 9 | | | |
| 10 | | | |
| 18 | | | |
| 19 | | | |
| 22 | | | |
| 26 | | | |

$\Sigma X=$ $\Sigma X^2 =$ $\Sigma(X-\overline{X})^2 =$

**16.** Compute the mean.

Mean =

**17.** Compute the variance.

Variance =

**18.** Compute the standard deviation.

Standard deviation =

Following are ratings by 9 married people of the degree of their spouse's attractiveness. Use these data to answer problems 19 through 21.

| $X$ | $X^2$ | $X-\overline{X}$ | $(X-\overline{X})^2$ |
|---|---|---|---|
| 500 | | | |
| 500 | | | |
| 523 | | | |
| 524 | | | |
| 550 | | | |
| 600 | | | |
| 625 | | | |
| 629 | | | |

$\Sigma X=$ $\Sigma X^2 =$ $\Sigma(X-\overline{X})^2 =$

**19.** Compute the mean.

Mean =

**20.** Compute the variance.

Variance =

**21.** Compute the standard deviation.

Standard deviation =

**Use the number of life change units for the following 12 subjects to answer problems 22 through 24.**

| $X$ | $X^2$ | $X-\overline{X}$ | $(X-\overline{X})^2$ |
|---|---|---|---|
| 150 | | | |
| 175 | | | |
| 375 | | | |
| 210 | | | |
| 400 | | | |
| 216 | | | |
| 300 | | | |
| 175 | | | |
| 374 | | | |
| 163 | | | |
| 263 | | | |
| 222 | | | |

$\Sigma X=$ $\Sigma X^2=$ $\Sigma(X-\overline{X})^2=$

**22.** Compute the mean.

Mean =

**23.** Compute the variance.

Variance =

**24.** Compute the standard deviation.

Standard deviation =

Use the following frequency distribution to answer problems 25 through 27.

| App Lim | F | $\bar{X}$ | FX | $X-\bar{X}$ | $(X-\bar{X})^2$ |
|---|---|---|---|---|---|
| 240 - 254 | 1 | | | | |
| 225 - 239 | 10 | | | | |
| 210 - 224 | 14 | | | | |
| 195 - 209 | 27 | | | | |
| 180 - 194 | 29 | | | | |
| 165 - 179 | 56 | | | | |
| 150 - 164 | 29 | | | | |
| 135 - 149 | 22 | | | | |
| 120 - 134 | 9 | | | | |
| 105 - 119 | 3 | | | | |

$\Sigma F=$ $\Sigma FX=$ $\Sigma(X-\bar{X})^2=$

**25.** Compute the mean.

Mean =

**26.** Compute variance.

Variance =

**27.** Compute the standard deviation.

Standard deviation =

A psychologist has completed a research project commissioned by a large corporation in which she compared the percentage of times employees responded to a memo sent through electronic mail with the percentage of times they responded to a memo sent through regular company mail. Use the following results to answer problems 28 through 31.

| Electronic Mail | Company Mail |
|---|---|
| Mean = 75 | Mean = 79 |
| Std. Dev. = 10 | Std. Dev. = 5 |

**28.** Given the above data, which type of mail produced the most consistent results? Why do you say that?

**29.** Of the two types of mail, which would you choose if you were the president of the company and were forced to choose only one? Why would you choose this type of mail?

30. Of the two types of mail, which is most likely to have the single highest response rate? Why do you think this?

31. Of the two types of mail, which is most likely to have the single lowest response rate? Why do you think this?

CHAPTER

# SCALED SCORES AND STANDARD SCORES

Suppose your statistics professor gives you a test, with the results producing a mean of 72 and a standard deviation of 8. Use this mean and standard deviation to answer questions 1 through 11.

1. What would be the new mean and new standard deviation if your professor decided to add 10 points to each score?

   New mean =

   New standard deviation =

2. What would be the new mean and new standard deviation if your professor decided to subtract 10 points from each score?

   New mean =

   New standard deviation =

3. What would be the new mean and new standard deviation if your professor decided to add 5 points to each score?

   New mean =

   New standard deviation =

4. What would be the new mean and new standard deviation if your professor decided to subtract 7.5 points from each score?

   New mean =

   New standard deviation =

5. What would be the new mean and new standard deviation if your professor decided to multiply each score by 2?

   New mean =

   New standard deviation =

6. What would be the new mean and new standard deviation if your professor decided to divide each score by 4?

   New mean =

   New standard deviation =

7. What would be the new mean and new standard deviation if your professor decided to multiply each score by 5?

   New mean =

   New standard deviation =

8. What would be the new mean and the new standard deviation if your professor decided to divide each score by 3?

New mean =

New standard deviation =

9. What would be the new mean and the new standard deviation if your professor decided to subtract 72 points from each score?

New mean =

New standard deviation =

10. What would be the new mean and the new standard deviation if your professor decided to divide each score by 12?

New mean =

New standard deviation =

11. What would be the new mean and the new standard deviation if your professor decided first to subtract 72 points from each score and then to divide the resulting scores by 12?

New mean =

New standard deviation =

Use the following reaction time scores to answer problems 12 through 14.

| $X$ | $X-\overline{X}$ | $(X-\overline{X})^2$ | $z$ |
|---|---|---|---|
| 223 | | | |
| 234 | | | |
| 234 | | | |
| 567 | | | |
| 236 | | | |
| 129 | | | |
| 438 | | | |
| 283 | | | |
| 444 | | | |
| 358 | | | |
| 445 | | | |

$\Sigma X=$ $\Sigma(X-\overline{X})^2=$

**12.** Compute the mean.

Mean =

**13.** Compute the standard deviation.

Standard deviation =

**14.** Convert each score in the distribution to a $z$ score.

Use the following achievement test scores to answer problems 15 through 17.

| $X$ | $X-\overline{X}$ | $(X-\overline{X})^2$ | $z$ |
|---|---|---|---|
| 99 | | | |
| 98 | | | |
| 98 | | | |
| 97 | | | |
| 95 | | | |
| 93 | | | |
| 92 | | | |
| 92 | | | |
| 91 | | | |
| 90 | | | |
| 90 | | | |
| 89 | | | |
| 88 | | | |
| 88 | | | |
| 86 | | | |
| 85 | | | |
| 83 | | | |

$\Sigma X=$ $\Sigma(X-\overline{X})^2=$

**15.** Compute the mean.

Mean =

**16.** Compute the standard deviation.

Standard deviation =

**17.** Convert each score in the distribution to a $z$ score.

Use the following video game scores for 10 elementary school children to answer problems 18 through 20.

| $X$ | $X-\bar{X}$ | $(X-\bar{X})^2$ | $z$ |
|---|---|---|---|
| 12 | | | |
| 14 | | | |
| 17 | | | |
| 18 | | | |
| 19 | | | |
| 23 | | | |
| 25 | | | |
| 28 | | | |
| 34 | | | |
| 59 | | | |

$\Sigma X=$ $\Sigma(X-\bar{X})^2=$

**18.** Compute the mean.

Mean =

**19.** Compute the standard deviation.

Standard deviation =

**20.** Convert each score in the distribution to a $z$ score.

The following data represent the number of minutes spent daydreaming each day by 15 subjects. Use these data to answer problems 21 through 23.

| $X$ | $X-\overline{X}$ | $(X-\overline{X})^2$ | $z$ |
|---|---|---|---|
| 12 | | | |
| 14 | | | |
| 16 | | | |
| 18 | | | |
| 19 | | | |
| 19 | | | |
| 20 | | | |
| 22 | | | |
| 23 | | | |
| 24 | | | |
| 35 | | | |
| 36 | | | |
| 39 | | | |
| 42 | | | |
| 47 | | | |

$\Sigma X=$ $\Sigma(X-\overline{X})^2=$

**21.** Compute the mean.

Mean =

**22.** Compute the standard deviation.

Standard deviation =

**23.** Convert each score in the distribution to a $z$ score.

Use the following data on the number of aggressive acts per hour for 13 preschool children to answer problems 24 through 26.

| $X$ | $X-\overline{X}$ | $(X-\overline{X})^2$ | $z$ |
|---|---|---|---|
| 1 | | | |
| 2 | | | |
| 2 | | | |
| 2 | | | |
| 2 | | | |
| 7 | | | |
| 7 | | | |
| 7 | | | |
| 8 | | | |
| 9 | | | |
| 10 | | | |
| 12 | | | |
| 15 | | | |

$\Sigma X=$ $\Sigma(X-\overline{X})^2=$

**24.** Compute the mean.

Mean =

**25.** Compute the standard deviation.

Standard deviation =

**26.** Convert each score in the distribution to a $z$ score.

The number of positive statements about their self-image was recorded for 15 patients. Use these data to answer problems 27 through 29.

| $X$ | $X-\overline{X}$ | $(X-\overline{X})^2$ | $z$ |
|---|---|---|---|
| 2 | | | |
| 2 | | | |
| 3 | | | |
| 4 | | | |
| 5 | | | |
| 5 | | | |
| 6 | | | |
| 7 | | | |
| 8 | | | |
| 9 | | | |
| 10 | | | |
| 18 | | | |
| 19 | | | |
| 22 | | | |
| 26 | | | |

$\Sigma X=$ $\Sigma(X-\overline{X})^2=$

**27.** Compute the mean.

Mean =

**28.** Compute the standard deviation.

Standard deviation =

**29.** Convert each score in the distribution to a $z$ score.

CHAPTER

# THE NORMAL CURVE

For problems 1 through 5, use Table Z in the textbook to find the area beyond the listed *z* score.

**1.** $z = 1.28$

Area beyond =

**2.** $z = .47$

Area beyond =

**3.** $z = 0$

Area beyond =

**4.** $z = -.29$

Area beyond =

**5.** $z = -2.56$

Area beyond =

| For problems 6 through 10, use Table Z to find the area between the mean and the $z$ score. |
|---|

**6.** $z = 2.04$

Area between mean and $z$ =

**7.** $z = 1.66$

Area between mean and $z$ =

**8.** $z = 0$

Area between mean and $z$ =

**9.** $z = -.89$

Area between mean and $z$ =

**10.** $z = -\ 1.35$

Area between mean and $z$ =

You have just finished measuring the speed at which all the rats in your colony run a complex maze. You have calculated the mean speed of your rat population to be 60 seconds and the standard deviation to be 10 seconds. Use this mean and standard deviation to answer problems 11 through 24.

**11.** What proportion of the rats takes *longer* than 75 seconds to run the maze?

Proportion longer than 75 =

**12.** What proportion of the rats takes *longer* than 60 seconds to run the maze?

Proportion longer than 60 =

**13.** What proportion of the rats takes *longer* than 54 seconds to run the maze?

Proportion longer than 54 =

**14.** What proportion of the rats takes *fewer* than 65 seconds to run the maze?

Proportion fewer than 65 =

**15.** What proportion of the rats takes *fewer* than 60 seconds to run the maze?

Proportion fewer than 60 =

**16.** What proportion of the rats takes *fewer* than 40 seconds to run the maze?

Proportion fewer than 40 =

**17.** What time, in seconds, represents the 90th percentile?

90th percentile =

**18.** What time represents the 50th percentile?

50th percentile =

**19.** What time represents the 15th percentile?

15th percentile =

**20.** What proportion of the rats in the colony runs the maze *between* 65 and 70 seconds?

Proportion between 65 and 70 =

**21.** What proportion of the rats runs the maze *between* 55 and 70 seconds?

Proportion between 55 and 70 =

**22.** What proportion of the rats in the colony runs the maze in *exactly* 70 seconds?

Proportion exactly 70 =

**23.** What proportion of the rats runs the maze in *exactly* 45 seconds?

Proportion exactly 45 =

**24.** What proportion of the rats runs the maze in *exactly* 47 seconds?

Proportion exactly 47 =

For your doctoral dissertation, you decide to study the depth-perception accuracy of 100,000 people. You ask them to pull some cords in order to line up two rods, side by side, from a distance of 6 meters. You then measure the distance between the two rods. Your results indicate that the average distance between the two rods is 100 millimeters, with a standard deviation of 20 millimeters. Use this mean and standard deviation to answer problems 25 through 35.

**25.** What proportion of people sets the rods *farther* than 125 millimeters apart?

Proportion farther than 125 =

**26.** What proportion of people sets the rods *farther* than 100 millimeters apart?

Proportion farther than 100 =

**27.** What proportion of people sets the rods *closer* than 120 millimeters apart?

Proportion closer than 120 =

**28.** What proportion of people sets the rods *closer* than 100 millimeters apart?

Proportion closer than 100 =

**29.** What proportion of people sets the rods *closer* than 70 millimeters apart?

Proportion closer than 70 =

**30.** What proportion of people sets the rods *between* 110 and 100 millimeters apart?

Proportion between 110 and 100 =

**31.** What proportion of people sets the rods *between* 90 and 75 millimeters apart?

Proportion between 90 and 75 =

**32.** What proportion of people sets the rods *exactly* 110 millimeters apart?

Proportion exactly 110 =

33. What proportion of people sets the rods *exactly* 90 millimeters apart?

Proportion exactly 90 =

34. What distance, in millimeters, represents the 73rd percentile?

73rd percentile =

35. What distance, in millimeters, represents the 33rd percentile?

33rd percentile =

36. In a normal distribution with a mean of 25, which of the following scores has the *least* likely chance of occurring? (Circle your choice.) Why did you pick this score?

27 20 31 26 25

37. In a normal distribution with a mean of 45, which of the following scores has the *most* likely chance of occurring? (Circle your choice.) Why did you pick this score?

46 23 39 49 52

38. In a normal distribution with a mean of 75, what proportion of the distribution has a score *greater* than 75? Explain your answer in one sentence.

CHAPTER

# CORRELATION

Use the following IQ scores from 10 different families to answer problems 1 through 7.

| Family | Mother | Father | Son | Daughter |
|---|---|---|---|---|
| 1 | 110 | 120 | 121 | 123 |
| 2 | 100 | 110 | 105 | 100 |
| 3 | 134 | 120 | 125 | 132 |
| 4 | 94 | 92 | 87 | 103 |
| 5 | 88 | 85 | 92 | 88 |
| 6 | 74 | 72 | 90 | 82 |
| 7 | 95 | 107 | 110 | 100 |
| 8 | 123 | 115 | 122 | 100 |
| 9 | 139 | 155 | 133 | 147 |
| 10 | 100 | 90 | 123 | 110 |

1. Draw a scatter plot using the Mother and Father scores for the 10 families.

2. Does the scatter plot you drew in problem 1 indicate a strong or a weak correlation? Please explain your answer in a few sentences.

3. Does the scatter plot you drew in problem 1 indicate a positive or negative correlation?

Correlation is __________ .

4. Draw a scatter plot using the Mother and Son scores for the 10 families.

5. Does the scatter plot you drew in problem 4 indicate a strong or a weak correlation? Please explain your answer in a few sentences.

6. Does the scatter plot you drew in problem 4 indicate a positive or negative correlation?

Correlation is __________ .

7. Draw a scatter plot using the Mother and Daughter scores for the 10 families.

8. Does the scatter plot you drew in problem 7 indicate a strong or a weak correlation? Please explain your answer in a few sentences.

9. Does the scatter plot you drew in problem 7 indicate a positive or negative correlation?

Correlation is __________ .

10. Draw a scatter plot using the Father and Son scores for the 10 families.

11. Does the scatter plot you drew in problem 10 indicate a strong or a weak correlation? Please explain your answer in a few sentences.

12. Does the scatter plot you drew in problem 10 indicate a positive or negative correlation?

Correlation is ________ .

**13.** Draw a scatter plot using the Father and Daughter scores for the 10 families.

**14.** Does the scatter plot you drew in problem 13 indicate a strong or a weak correlation? Please explain your answer in a few sentences.

**15.** Does the scatter plot you drew in problem 13 indicate a positive or negative correlation?

Correlation is __________ .

**16.** Draw a scatter plot using the Son and Daughter scores for the 10 families.

**17.** Does the scatter plot you drew in problem 16 indicate a strong or a weak correlation? Please explain your answer in a few sentences.

**18.** Does the scatter plot you drew in problem 16 indicate a positive or negative correlation?

Correlation is __________ .

The following data were collected from 6 male college students: IQ score, SAT Verbal score, grade point average, birth order, and shoe size. Use these data in problems 19 through 32.

| Student | IQ | SAT | GPA | Birth Order | Shoe Size |
|---|---|---|---|---|---|
| M. J. | 123 | 540 | 3.7 | 1 | 7.5 |
| N. C. | 105 | 430 | 2.6 | 5 | 9.5 |
| O. P. | 95 | 320 | 1.9 | 3 | 11.0 |
| M. T. | 134 | 690 | 3.9 | 1 | 7.0 |
| A. D. | 123 | 550 | 3.0 | 3 | 13.0 |
| J. L. | 110 | 530 | 2.9 | 2 | 6.0 |
| Mean | 115 | 510 | 3.0 | 2.5 | 9.0 |
| $\Sigma X^2$ | 80364 | 1638400 | 56.68 | 49 | 521.5 |
| $S$ | 13 | 113.871 | .668 | 1.384 | 2.432 |

**19.** Use the $z$-score formula to compute the correlation coefficient between IQ score and SAT score for these 6 students.

| IQ Score $X$ | $z_X$ | SAT Score $Y$ | $z_Y$ | $z_X \cdot z_Y$ |
|---|---|---|---|---|

$r$ =

**20.** Compute the coefficient of determination for problem 19. Given this coefficient of determination, explain in words the relationship between IQ score and SAT score for these 6 students.

$r^2$ =

**21.** Use the $z$-score formula to compute the correlation coefficient between GPA and birth order for these 6 students.

| GPA $X$ | $z_X$ | Birth Order $Y$ | $z_Y$ | $z_X z_Y$ |
|---|---|---|---|---|

$r$ =

**22.** Compute the coefficient of determination for problem 21. Given this coefficient of determination, explain in words the GPA and birth order relationship for these 6 students.

$r^2$ =

**23.** Use the computational formula to compute the correlation coefficient between birth order and IQ score for these 6 students.

| Birth Order $X$ | IQ Score $Y$ | $X \cdot Y$ |
|---|---|---|

$r =$

**24.** Compute the coefficient of determination for problem 23. Given this coefficient of determination, explain in words the relationship between birth order and IQ score these 6 students.

$r^2 =$

**25.** Use the computational formula to compute the correlation coefficient between IQ score and shoe size for these 6 students.

| IQ Score $X$ | Shoe Size $Y$ | $X \cdot Y$ |
| --- | --- | --- |

$r =$

**26.** Compute the coefficient of determination for problem 25. Given this coefficient of determination, explain in words the relationship between IQ score and shoe size for these 6 students.

$r^2 =$

**27.** Use the computational formula to compute the correlation coefficient between GPA and SAT score for these 6 students.

| GPA $X$ | SAT Score $Y$ | $X \cdot Y$ |
|---|---|---|

$r$ =

**28.** Compute the coefficient of determination for problem 27. Given this coefficient of determination, explain in words the relationship between GPA and SAT score for these 6 students.

$r^2$ =

CHAPTER

# REGRESSION

A psychologist has found a correlation between the amount of popcorn a subject eats during a movie and the number of days since the subject's last diet. The means, standard deviations, and the correlation coefficient follow. Use this information to answer problems 1 through 6.

| Number of popcorn pieces $X$ | Days since last diet $Y$ |
|---|---|
| $\overline{X} = 50$ | $\overline{Y} = 25$ |
| $S_X = 15$ | $S_Y = 5$ |

$r = .60$

1. How many pieces of popcorn do you predict a subject will eat if it has been 10 days since the subject's last diet?

$X' =$

2. How many pieces of popcorn do you predict a subject will eat if it has been 20 days since the subject's last diet?

$X' =$

3. How many pieces of popcorn do you predict a subject will eat if it has been 35 days since the subject's last diet?

$X' =$

4. How many days do you predict it has been since the subject's last diet if the subject eats 60 pieces of popcorn?

$Y' =$

5. How many days do you predict it has been since the subject's last diet if the subject eats 65 pieces of popcorn?

$Y' =$

6. How many days do you predict it has been since the subject's last diet if the subject eats 50 pieces of popcorn?

$Y' =$

A psychologist conducting research on the effect of alcohol on maze learning in rats has found that the time taken by a drunken rat to run a maze is correlated with the time taken by the rat when sober. The data from her research, recorded in seconds, follow. Use this information to answer problems 7 through 12.

| Time to run maze sober $X$ | Time to run maze drunk $Y$ |
|---|---|
| $\bar{X} = 90$ | $\bar{Y} = 125$ |
| $S_X = 25$ | $S_Y = 35$ |

$r = .75$

7. Predict how long it will take a rat to run the maze when it is sober if it takes the rat 100 seconds to run the maze when it is drunk.

$X' =$

**8.** Predict how long it will take a rat to run the maze when it is sober if it takes the rat 155 seconds to run the maze when it is drunk.

$X' =$

**9.** Predict how long it will take a rat to run the maze when it is sober if it takes the rat 125 seconds to run the maze when it is drunk.

$X' =$

**10.** Predict how long it will take a rat to run the maze when it is drunk if it takes the rat 120 seconds to run the maze when it is sober.

$Y' =$

**11.** Predict how long it will take a rat to run the maze when it is drunk if it takes the rat 110 seconds to run the maze when it is sober.

$Y' =$

**12.** Predict how long it will take a rat to run the maze when it is drunk if it takes the rat 90 seconds to run the maze when it is sober.

$Y' =$

> A cognitive psychologist who has just completed standardization work on a new Verbal IQ Test (VIQT) discovered that the new test is negatively correlated with the Psychic Ability Test (PAT). The means and standard deviations for the two tests, as well as the correlation between the two tests, follow. Use these data to answer problems 13 through 18.

| Verbal IQ Test (VIQT) $X$ | Psychic Ability Test (PAT) $Y$ |
|---|---|
| $\overline{X} = 200$ | $\overline{Y} = 20$ |
| $s_X = 20$ | $s_Y = 5$ |

$r = -.45$

**13.** What score do you predict a person will get on the VIQT if she gets a 15 on the PAT?

$X' =$

**14.** What score do you predict a person will get on the VIQT if he gets a 5 on the PAT?

$X' =$

**15.** What score do you predict a person will get on the VIQT if she gets a 20 on the PAT?

$X' =$

**16.** What score do you predict a person will get on the PAT if he gets a 210 on the VIQT?

$Y' =$

**17.** What score do you predict a person will get on the PAT if she gets a 240 on the VIQT?

$Y' =$

**18.** What score do you predict a person will get on the PAT if he gets a 200 on the VIQT?

$Y' =$

The MMPFI, Married Men's Personality and Faithfulness Inventory, has been used for years by marriage and family counselors in your clinic. The new version of the test, the MMPFI-2, has just been released. Use the following data to answer questions 19 through 24.

| MMPFI $X$ | MMPFI-2 $Y$ |
|---|---|
| $\overline{X} = 25$ | $\overline{Y} = 30$ |
| $S_X = 5$ | $S_Y = 6$ |

$r = .80$

**19.** What score would you predict on the MMPFI-2 for a patient who scored 20 on the MMPFI?

$Y' =$

**20.** What score would you predict on the MMPFI-2 for a patient who scored 15 on the MMPFI?

$Y' =$

**21.** What score would you predict on the MMPFI-2 for a patient who scored 25 on the MMPFI?

$Y' =$

**22.** What score would you predict on the MMPFI for a patient who scored 20 on the MMPFI-2?

$X' =$

**23.** What score would you predict on the MMPFI for a patient who scored 24 on the MMPFI-2?

$X' =$

**24.** What score would you predict on the MMPFI for a patient who scored 30 on the MMPFI-2?

$X' =$

CHAPTER 10

# PROBABILITY THEORY AND SAMPLING

The following information is a breakdown of the various disorders suffered by 2000 patients with major mental disorders at State Mental Hospital. Use this information to answer problems 1 through 10, which ask questions about patients selected at random from the pool of hospital patients.

| Disorder | Number of patients |
|---|---|
| Substance abuse | 600 |
| Mood | |
| Major depression | 250 |
| Bipolar | 200 |
| Schizophrenia | |
| Paranoid | 500 |
| Catatonic | 50 |
| Undifferentiated | 200 |
| Disorganized | 100 |
| Other | 100 |
| Total | 2000 |

1. What is the probability of a patient being diagnosed as having a *substance abuse* disorder?

   Probability =

2. What is the probability of a patient being diagnosed as having *major depression*?

   Probability =

3. What is the probability of a patient being diagnosed as having a *bipolar disorder*?

   Probability =

4. What is the probability of a patient being diagnosed as having *paranoid schizophrenia*?

   Probability =

5. What is the probability of a patient being diagnosed as having *catatonic schizophrenia*?

   Probability =

6. What is the probability of a patient being diagnosed as having *undifferentiated schizophrenia*?

Probability =

7. What is the probability of a patient being diagnosed as having *disorganized schizophrenia*?

Probability =

8. What is the probability of a patient being diagnosed as having a disorder listed as *other*?

Probability =

9. What is the probability of a patient being diagnosed as having a *mood* disorder?

Probability =

10. What is the probability of a patient being diagnosed as having *schizophrenia*?

Probability =

Following is a list of demographic data from the psychology department at a large university. Listed are several demographic categories, as well as the proportion of faculty members fitting into the categories. *Assuming that the major categories are mutually exclusive*, use this information to answer questions 11 through 30.

| Category | Proportion |
|---|---|
| Ethnic background | |
| White | .70 |
| Hispanic | .15 |
| Asian | .09 |
| Black | .05 |
| Native American | .01 |
| Gender | |
| Male | .55 |
| Female | .45 |
| Highest degree | |
| Ph.D. | .75 |
| M.A. | .20 |
| B.A. | .04 |
| A.A. | .01 |
| Area of specialization | |
| Clinical/personality | .40 |
| Cognitive | .25 |
| Social | .10 |
| Behavioral | .10 |
| Physiological | .05 |
| Sensation/perception | .05 |
| Developmental | .05 |

**11.** What is the probability of a faculty member being female and having a Ph.D.?

Probability =

12. What is the probability of a faculty member being a Ph.D. and a social psychologist?

Probability =

13. What is the probability of a faculty member being a black clinical/personality psychologist?

Probability =

14. What is the probability of a faculty member being female and black?

Probability =

15. What is the probability of a faculty member being Asian and a developmental psychologist?

Probability =

16. What is the probability of a faculty member being a behaviorist with a Ph.D.?

Probability =

17. What is the probability of a faculty member being a female developmental psychologist with an A.A. degree?

Probability =

18. What is the probability of a faculty member being a Hispanic male cognitive psychologist?

Probability =

19. What is the probability of a faculty member being a Native American male behaviorist with a Ph.D. degree?

Probability =

20. What is the probability of a faculty member being female?

Probability =

21. What is the probability of a faculty member having either an M.A. or an A.A. degree?

Probability =

22. What is the probability of a faculty member being either a physiological psychologist or a sensation/perception psychologist?

Probability =

23. What is the probability of a faculty member being black, Native American, Asian, or Hispanic?

Probability =

24. What is the probability of a faculty member being either female or male?

Probability =

25. What is the probability of a faculty member being either a developmental psychologist or a sensation/perception psychologist?

Probability =

26. What is the probability of a faculty member being Hispanic?

Probability =

27. What is the probability of a faculty member being either a black male or a female with a Ph.D.?

Probability =

28. What is the probability of a faculty member being a male with a Ph.D. or a male with an M.A. degree?

Probability =

29. What is the probability of a faculty member being a female cognitive psychologist with a Ph.D.?

Probability =

30. What is the probability of a faculty member being a male with a Ph.D.?

Probability =

> An industrial psychologist was employed by a company to help solve a severe problem with interpersonal relations in a major department. The first thing she did was to administer a personality test to the department supervisors. Use the following test score data to answer problems 31 through 34.

| Personality test scores |
|---|
| 125 |
| 120 |
| 100 |
| 85 |
| 110 |

31. Compute the variance of the sample of personality test scores.

$S^2$ =

32. Compute the standard deviation of the sample of personality test scores.

S =

**33.** Compute the estimate of the population variance using the sample of personality test scores.

est. $\sigma^2$ =

**34.** Compute the estimate of the population standard deviation using the sample of personality test scores.

est. $\sigma$ =

| The industrial psychologist gave another test of social affiliation to random workers in the department to determine how much they enjoyed being with others. Use the following social affiliation scores to answer problems 35 through 38. |
|---|

| Social affiliation scores |
|---|
| 20 |
| 25 |
| 30 |
| 25 |
| 35 |
| 30 |
| 20 |
| 15 |
| 25 |
| 30 |

**35.** Compute the variance of the sample of social affiliation scores.

$S^2$ =

**36.** Compute the standard deviation of the sample of social affiliation scores.

$S =$

**37.** Compute the estimate of the population variance using the sample of social affiliation scores.

est. $\sigma^2 =$

**38.** Compute the estimate of the population standard deviation using the sample of social affiliation scores.

est. $\sigma =$

The following scores, from an experiment on conformity, are the number of times a subject conformed to the majority when the majority was wrong. Use these data to answer problems 39 through 42.

| Number of conforming answers |
|---|
| 5 |
| 4 |
| 3 |
| 7 |
| 5 |
| 4 |
| 3 |
| 10 |
| 7 |
| 4 |

39. Compute the variance of the sample of conformity scores.

$S^2 =$

40. Compute the standard deviation of the sample of conformity scores.

$S =$

41. Compute the estimate of the population variance using the sample of conformity scores.

est. $\sigma^2 =$

42. Compute the estimate of the population standard deviation using the sample of conformity scores.

est. $\sigma =$

A psychologist studying the sizes of groups formed by teenagers at a high school dance recorded the following data for eight different groups. Use these data to answer problems 43 to 46.

| Group Size |
| --- |
| 2 |
| 4 |
| 3 |
| 7 |
| 5 |
| 8 |
| 7 |
| 3 |

**43.** Compute the variance of this sample of scores.

$S^2$ =

**44.** Compute the standard deviation of this sample of scores.

$S$ =

**45.** Compute the estimate of the population variance using this sample.

est. $\sigma^2$ =

**46.** Compute the estimate of the population standard deviation using this sample.

est. $\sigma$ =

# CHAPTER 11

# EXPERIMENTAL DESIGN

1. A scientific research ____________________ is a prediction predominantly based on a scientific theory or body of knowledge.

2. Why would the following hypothesis be unacceptable as a scientific hypothesis?

   "Psychologists who conduct research on extra sensory perception (ESP) and who do not believe in ESP will tend to get negative results because their negative thoughts will interfere with the ESP of their subjects."

3. ____________________ variables are manipulated by the experimenter and applied to the subject in order to determine their effect on the subject's behavior.

4. ______________________ variables are used to assess or measure the subject's behavior.

5. ______________________ variables describe subjects' characteristics or attributes that cannot be manipulated by the experimenter.

A social psychologist is investigating blushing behavior in male and female subjects. The experimenter randomly assigns 20 male and female subjects to either of two conditions: one in which they view pictures of clothed male and female models and another in which they view pictures of nude male and female models. Because blushing is the result of increased blood flow to facial skin, the psychologist can measure blushing behavior by measuring the change in the temperature of the subjects' skin. The psychologist expects an increase in the amount of blushing when the subject views the nude photos as opposed to the clothed photos. The psychologist also expects a difference between the blushing behavior of males and females. Use this hypothetical research project to answer problems 6 through 9.

6. Is this blushing research an experiment? (Circle Yes or No.) Also explain the reasons for your answer.

Yes No

Why?

7. In the blushing research, what is the independent variable? If there is no independent variable, answer *none*. Also explain why you gave this answer.

Independent Variable:

Why?

8. In the blushing research, what is the dependent variable? If there is no dependent variable, answer *none*. Explain why you gave this answer.

Dependent Variable:

Why?

9. In the blushing research, identify any subject variables. If there are no subject variables, write *none*. Explain why you believe that there are or are not subject variables in this experiment.

Subject Variables:

Why?

Depression is a major disorder that affects one of four women and one of eight men at some time during their lives. A psychologist is interested in seeing if there is a relationship between the number of episodes of depression and the gender of the patient. Use this hypothetical research project to answer problems 10 through 13.

**10.** Is the depression research an experiment? (Circle Yes or No.) Explain the reasons for your answer.

Yes No

Why?

**11.** In the depression research, what is the independent variable? If there is no independent variable, answer *none*. Explain why you gave this answer.

Independent Variable:

Why?

**12.** In the depression research, what is the dependent variable? If there is no dependent variable, answer *none*. Explain why you gave this answer.

Dependent Variable:

Why?

**13.** In the depression research, identify any subject variables. If there are no subject variables, write *none*. Explain why you believe that there are or are not subject variables.

Subject Variables:

Why?

A cognitive psychologist is interested in subjects' memories for complex moving patterns. She randomly assigns subjects to three groups. Each group watches as the psychologist moves a complex pattern of shapes in one of three ways--rotating the pattern, moving the pattern back and forth horizontally, and moving the pattern up and down vertically. Twenty-four hours later, the psychologist records the number of pattern elements each subject remembers. Use this hypothetical research project to answer problems 18 through 21.

**14.** Is this research on pattern memory an experiment? (Circle Yes or No.) Explain the reasons for your answer.

Yes No

Why?

**15.** In the research on pattern memory, what is the independent variable? If there is no independent variable, answer *none*. Explain why you gave this answer.

Independent Variable:

Why?

**16.** In the research on pattern memory, what is the dependent variable? If there is no dependent variable, answer *none*. Explain why you gave this answer.

Dependent Variable:

Why?

**17.** In the research on pattern memory, identify any subject variables. If there are no subject variables, write *none*. Explain why you believe that there are or are not subject variables.

Subject Variables:

Why?

Problems 26 through 30 describe different experiments. Each of these experiments has either a one-group design, a completely randomized design, or a completely randomized factorial design. Write down which design is used in each experiment and why you think you are correct.

**18.** A researcher is conducting an experiment with two independent variables. Each independent variable has three levels.

Type of Design:

Why?

**19.** A researcher is conducting an experiment that has one independent variable with three levels.

Type of design:

Why?

**20.** A researcher is conducting an experiment in which she chooses to compare the mean of one experimental sample to the known mean of the population.

Type of design:

Why?

**21.** A researcher is conducting an experiment comparing an experimental group to a control group.

Type of design:

Why?

**22.** A researcher is conducting an experiment with three independent variables. The first independent variable has two levels, the second independent variable has three levels, and the third independent variable has four levels.

Type of design:

Why?

CHAPTER 

# T TESTS

An educational psychologist is interested in seeing if reentry students, students who have not attended school for more than four years in order to work or raise a family, are more motivated to achieve higher grades than students who have not taken a break in their education. He knows that the grade point average for students at his college who have not interrupted their education, the population value, is equal to 2.9. Use the following sample of GPAs for 10 reentry students to answer problems 1 through 8.

| Reentry student | GPA |
|---|---|
| R. J. | 3.7 |
| B. N. | 2.6 |
| M. V. | 3.1 |
| T. D. | 3.0 |
| U. C. | 3.1 |
| P. T. | 3.8 |
| M. M. | 2.6 |
| N. B. | 3.4 |
| M. A. | 3.0 |
| G. P. | 2.8 |

1. Compute the mean GPA for the reentry students.

$\overline{GPA}$ =

2. Compute the standard deviation for the reentry students.

$S$ =

3. Compute the estimate of the population standard deviation using the sample of reentry students.

est. $\sigma$ =

4. Compute the estimate of the standard error of the mean.

est. $\sigma_{\overline{X}}$ =

5. Compute $t$.

$t$ =

6. What are the degrees of freedom for this $t$?

$df$ =

7. Look up this $t$ in Table T and determine if it is significant for a *one-tailed* test.

   Critical value =

   Significant? Yes No

8. Given the $t$ test you have just conducted, what are your conclusions about the GPA of reentry students?

As a psychologist interested in eating behavior, you decide to determine if scary movies cause people to eat more popcorn than musicals. You randomly assign 10 subjects to a group that watches a scary movie (*Psycho*) and another 10 subjects to a group that watches a musical (*The Sound of Music*). At the beginning of the movie, you give each subject a tub of 84 pieces of popcorn and tell the subjects not to share their popcorn with anyone. At the end of the movie, you measure the number of pieces of popcorn eaten by each subject. Use the following data for this hypothetical experiment to answer problems 9 through 16.

| Scary $X_1$ | Musical $X_2$ |
|---|---|
| 45 | 32 |
| 67 | 38 |
| 69 | 33 |
| 56 | 49 |
| 73 | 44 |
| 56 | 60 |
| 63 | 48 |
| 84 | 36 |
| 49 | 23 |
| 56 | 39 |

**9.** Compute the mean pieces of popcorn for each of the two samples.

$\overline{X}_1 =$

$\overline{X}_2 =$

**10.** Compute the standard deviation for each of the two samples.

$s_1 =$

$s_2 =$

**11.** Compute the estimate of the standard error of the mean for each sample.

est. $\sigma_{\overline{X}_1} =$

est. $\sigma_{\overline{X}_2} =$

**12.** Compute the estimate of the standard error of the difference.

est. $\sigma_{Diff} =$

**13.** Compute $t$.

$t =$

**14.** What are the degrees of freedom for this $t$?

$df$ =

**15.** Look up this $t$ in Table T and determine if it is significant for a *two-tailed* test.

Critical value =

Significant? Yes No

**16.** Given the $t$ test you have just conducted, what are your conclusions about the difference between the number of pieces of popcorn eaten by subjects viewing a scary movie versus a musical?

An educational psychologist is interested in the ability of preschool children to solve math story problems. He wants to see if the method of presentation--either as verbal story problems or as visual story problems--makes a difference in preschoolers' abilities to solve the problems correctly. In an example of the verbal condition, a child might be asked, "Two birds are sitting on a fence; two more birds fly down and join them. How many birds are on the fence altogether?" In an example of the nonverbal, visual equivalent of this problem, the experimenter might present the child with a picture of two birds on the fence with two birds in the process of landing on the fence and then ask the child, "How many birds are on the fence altogether?" In both conditions, the the child responds orally. Use the following data, which consist of the number of correct problems out of 10, to answer problems 17 through 26.

| Child | Verbal | Nonverbal | $D$ | $D^2$ |
|---|---|---|---|---|
| C. J. | 3 | 6 | | |
| F. K. | 5 | 8 | | |
| M. O. | 7 | 9 | | |
| I. M. | 4 | 8 | | |
| G. G. | 2 | 4 | | |
| K. T. | 1 | 1 | | |
| B. W. | 4 | 3 | | |
| M. B. | 2 | 8 | | |

**17.** Compute the differences.

**18.** Compute the differences squared.

**19.** Compute the sum of the differences.

$\Sigma D =$

**20.** Compute the sum of the differences squared.

$\Sigma D^2$ =

**21.** Compute the mean of the differences.

$\overline{D}$ =

**22.** Compute the estimate of the standard error of the difference.

est. $\sigma_{Diff}$ =

**23.** Compute the *t*.

$t$ =

**24.** What are the degrees of freedom for this *t*?

$df$ =

**25.** Look up this *t* in Table T and determine if it is significant for a *two-tailed* test.

Critical value =

Significant? Yes No

**26.** Given the *t* test you have just conducted, what are your conclusions about the difference between a preschooler's ability to solve simple math problems presented either verbally or nonverbally?

A human factors psychologist working for a hammer manufacturer is asked to evaluate the effectiveness of a new hammer design over the traditional hammer design. To test the new design, he measures the number of missed nails out of 100 tries. He knows from previous research that the traditional hammer has a population mean of 7 misses, with a population standard deviation of 2 misses. In his sample of 16 men using the new hammer design, he records a sample mean of 5 misses. Use this information to answer problems 27 through 31.

**27.** Compute the standard error of the mean.

$\sigma_{\overline{X}} =$

**28.** Compute a *t* for these data.

$t =$

**29.** What are the degrees of freedom for this *t*?

*df* =

**30.** Look up this *t* in Table T and determine if it is significant for a *two-tailed* test.

Table T value =

Significant? Yes No

**31.** Given the *t* test you have just conducted, what are your conclusions about the difference between the two hammer designs?

Of all the skills performed by football players, kicking requires the most concentration. A sports psychologist is interested in seeing if self-hypnosis can increase concentration and therefore, improve the kicking distance of a sample of 10 punters. She asks each punter to kick the ball five times and records the farthest of their five kicks. The following day, she teaches the punters a simple self-hypnosis ritual that focuses their concentration on their punting. The next day, the psychologist tells the punters to use self-hypnosis immediately before each of their five kicks. After each punter completes his five kicks, she records the farthest one. The means and standard deviations, in yards, for the pre- and post-hyposis conditions, as well as the correlation between the two, are shown below. Use this information to answer problems 32 through 37.

| Normal $X_1$ | Self-hypnotized $X_2$ |
| --- | --- |
| $\overline{X}_1 = 47$ | $\overline{X}_2 = 52$ |
| $s_1 = 6$ | $s_2 = 9$ |

$r = .80$

**32.** Compute the estimate of the standard error of the mean for the normal kicks as well as the estimate of the standard error of the mean for the self-hypnotized kicks.

est. $\sigma_{\overline{X}_1} =$

est. $\sigma_{\overline{X}_2} =$

33. Compute the estimate of the standard error of the difference for the two samples.

est. $\sigma_{Diff}$ =

34. Compute a *t*.

*t* =

35. What are the degrees of freedom for this *t*?

*df* =

36. Look up this *t* in Table T and determine if it is significant for a *two-tailed* test.

Critical value =

Significant? Yes No

37. Given the *t* test you have just conducted, what are your conclusions about the difference between the normal kicks and the kicks performed after self-hypnosis?

CHAPTER 13

# ONE-WAY ANALYSIS OF VARIANCE

A drug company has recently released a new antipsychotic drug that it claims decreases the delusions associated with schizophrenia. A psychiatrist and a psychobiologist are interested in determining the effective dosage of the new drug. They randomly assign 5 schizophrenics to one of three dosage conditions: A, B, and C. The average daily number of expressed delusions for each subject is shown in the table below. Use these data to answer problems 1 through 12.

Dosage condition

| A | | B | | C | |
|---|---|---|---|---|---|
| $X_1$ | $X_1^2$ | $X_2$ | $X_1^2$ | $X_3$ | $X_1^2$ |
| 22 | | 16 | | 5 | |
| 20 | | 14 | | 7 | |
| 19 | | 12 | | 4 | |
| 17 | | 17 | | 9 | |
| 27 | | 15 | | 8 | |

$\Sigma X_1 =$ $\Sigma X_2 =$ $\Sigma X_3 =$

$\Sigma X_1^2 =$ $\Sigma X_2^2 =$ $\Sigma X_3^2 =$

$\Sigma\Sigma X =$ $\Sigma\Sigma X^2 =$

1. Compute the sum of squares between groups.

$SS_{bg}$ =

2. Compute the sum of squares total.

$SS_{total}$ =

3. Compute the sum of squares within groups.

$SS_{wg}$ =

4. Compute the degrees of freedom between groups.

$df_{bg}$ =

5. Compute the degrees of freedom within groups.

$df_{wg}$ =

6. Compute the degrees of freedom total.

$df_{total}$ =

7. Compute the mean square between groups.

$MS_{bg}$ =

8. Compute the mean square within groups.

$MS_{wg}$ =

9. Compute $F$.

$F$ =

**10.** Assuming a *two-tailed* hypothesis, is the value of $F$ computed in problem 9 significant at the .05 level?

Critical value =

Significant? Yes No

**11.** If the value of $F$ computed in problem 9 is significant, compute the *HSD*.

*HSD* =

**12.** Given the values computed in the preceding problems, what are your conclusions about dosages A, B, and C? If you were the psychiatrist, which dosage level would you prescribe for your patients?

A group of psychologists is interested in seeing if self-help audio tapes can improve a subject's memory. They randomly assign 8 people to one of three conditions. In the audible condition, subjects listen to a tape that discusses a well-known mnemonic technique at a loudness level that is clearly audible. In the subliminal condition, subjects listen to a tape with the same message, but it is played at such a low volume that none of the subjects can consciously hear the message. In the control condition, subjects "listen," but the tape is blank and has no message. The following day, the subjects return and are presented with a list of words. After a five-minute break, they are given a test to see how many words they remember. The dependent variable in this experiment is the score received on the memory test. Use the following data to answer problems 13 through 24.

Type of Tape

| Audible $X_1$ | $X_1^2$ | Subliminal $X_2$ | $X_2^2$ | Control $X_3$ | $X_3^2$ |
|---|---|---|---|---|---|
| 55 | | 32 | | 30 | |
| 62 | | 30 | | 29 | |
| 49 | | 28 | | 26 | |
| 55 | | 31 | | 33 | |
| 61 | | 33 | | 28 | |

$\Sigma X_1 =$ $\Sigma X_2 =$ $\Sigma X_3 =$

$\Sigma X_1^2 =$ $\Sigma X_2^2 =$ $\Sigma X_3^2 =$

$\Sigma\Sigma X =$ $\Sigma\Sigma X^2 =$

**13.** Compute the sum of squares between groups.

$SS_{bg} =$

14. Compute the sum of squares total.

$SS_{\text{total}}$ =

15. Compute the sum of squares within groups.

$SS_{\text{wg}}$ =

16. Compute the degrees of freedom between groups.

$df_{\text{bg}}$ =

17. Compute the degrees of freedom within groups.

$df_{\text{wg}}$ =

18. Compute the degrees of freedom total.

$df_{\text{total}}$ =

**19.** Compute mean square between groups.

$MS_{bg}$ =

**20.** Compute the mean square within groups.

$MS_{wg}$ =

**21.** Compute *F*.

*F* =

**22.** Assuming a *two-tailed* hypothesis, is the value of *F* computed in problem 21 significant at the .05 level?

Critical value =

Significant? Yes No

**23.** If the value of *F* computed in problem 21 is significant, compute the *HSD*.

*HSD* =

**24.** Given the preceding values, what are your conclusions about the effect of the listening tapes on subjects' memories?

A cognitive psychologist has noticed that complex objects take longer to recognize than simple objects. He believes this difference is due to the fact that recognition of complex objects requires more eye movements than simple objects and therefore, take more time to process. To test his hypothesis, he monitors the eye movements of four groups of 6 subjects in order to count the number of fixations they make when viewing one of four different shapes: a triangle, a pentagon, an octagon, and a dodecagon (a 12-sided polygon). The number of fixations used by each subject to correctly identify each object is shown in the table below. Use these data to answer problems 25 through 36.

Type of Shape

| Triangle | | Pentagon | | Octagon | | Dodecagon | |
|---|---|---|---|---|---|---|---|
| $X_1$ | $X_1^2$ | $X_2$ | $X_2^2$ | $X_3$ | $X_3^2$ | $X_4$ | $X_4^2$ |
| 1 | | 2 | | 2 | | 5 | |
| 2 | | 3 | | 4 | | 6 | |
| 1 | | 4 | | 5 | | 6 | |
| 2 | | 4 | | 5 | | 6 | |
| 2 | | 3 | | 4 | | 5 | |
| 1 | | 3 | | 5 | | 7 | |

$\Sigma X_1 =$ $\Sigma X_2 =$ $\Sigma X_3 =$ $\Sigma X_4 =$

$X_1^2 =$ $\Sigma X_2^2 =$ $\Sigma X_3^2 =$ $\Sigma X_4^2 =$

$\Sigma\Sigma X =$ $\Sigma\Sigma X^2 =$

**25.** Compute the sum of squares between groups.

$SS_{bg}$ =

**26.** Compute the sum of squares total.

$SS_{total}$ =

**27.** Compute the sum of squares within groups.

$SS_{wg}$ =

**28.** Compute the degrees of freedom between groups.

$df_{bg}$ =

**29.** Compute the degrees of freedom within groups.

$df_{wg}$ =

30. Compute the degrees of freedom total.

$df_{total}$ =

31. Compute the mean square between groups.

$MS_{bg}$ =

32. Compute the mean square within groups.

$MS_{wg}$ =

33. Compute *F*.

*F* =

34. Assuming a *two-tailed* hypothesis, is the value of *F* computed in problem 33 significant at the .05 level?

Critical value =

Significant? Yes No

35. If the value of $F$ computed in problem 33 is significant, compute the $HSD$.

$HSD =$

36. Given the preceding values, what are your conclusions about the different samples in this experiment?

# CHAPTER 14

# FACTORIAL ANALYSIS OF VARIANCE

A psychologist studying maze learning in rats decides to see if diet or the environment in which rats are raised affects the ability of rats to learn a maze. He randomly assigns rats to either a high protein diet or a low protein diet, and he raises the rats either alone in individual cages or in colonies with large numbers of other rats. Thus, one independent variable in this project is the type of diet, while the other independent variable is the type of environment in which the rats are raised. The dependent variable is the number of trials it takes each rat to learn a complex maze. Use the following data to answer problems 1 through 22.

| | Diet | | |
|---|---|---|---|
| | Low protein | High protein | Row Total |
| Raised alone | 67 | 43 | |
| | 68 | 32 | |
| | 70 | 38 | |
| | 62 | 32 | |
| | 65 | 37 | |
| | $\Sigma X$ = | $\Sigma X$ = | |
| | $\Sigma X^2$ = | $\Sigma X^2$ = | |
| Raised in a colony | 43 | 28 | |
| | 44 | 24 | |
| | 45 | 27 | |
| | 39 | 25 | |
| | 44 | 29 | |
| | $\Sigma X$ = | $\Sigma X$ = | |
| | $\Sigma X^2$ = | $\Sigma X^2$ = | |
| | Column Total = | Column Total = | |
| | $\Sigma\Sigma X$ = | $\Sigma\Sigma X^2$ = | |

1. Compute the sum of squares total.

   $SS_{\text{total}}$ =

2. Compute the sum of squares within groups.

   $SS_{\text{wg}}$ =

3. Compute the sum of squares of the rows.

$SS_r$ =

4. Compute the sum of squares of the columns.

$SS_c$ =

5. Compute the sum of squares of the interaction.

$SS_{rXc}$ =

6. Compute the degrees of freedom total.

$df_{total}$ =

7. Compute the degrees of freedom within groups.

$df_{wg}$ =

8. Compute the degrees of freedom of the rows.

$df_r$ =

9. Compute the degrees of freedom of the columns.

$df_c$ =

10. Compute the degrees of freedom of the interaction.

$df_{rXc}$ =

11. Compute the mean square within groups.

$MS_{wg}$ =

12. Compute the mean square of the rows.

$MS_r$ =

**13.** Compute the mean square of the columns.

$MS_c$ =

**14.** Compute the mean square of the interaction.

$MS_{rXc}$ =

**15.** Compute the $F$ ratio for the rows.

$F_r$ =

**16.** Compute the $F$ ratio for the columns.

$F_c$ =

**17.** Compute the $F$ ratio for the interaction.

$F_{rXc}$ =

18. Is the *F* ratio for the rows significant for a *two-tailed* test?

*df* = ( , )

Critical value =

Significant? Yes No

19. Is the *F* ratio for the columns significant for a *two-tailed* test?

*df* = ( , )

Critical value =

Significant? Yes No

20. Is the *F* ratio for the interaction significant for a *two-tailed* test?

*df* = ( , )

Critical value =

Significant? Yes No

21. Use the space below to graph the means for each cell in order to illustrate the interaction.

22. Given the significance of the *F* ratios you just computed, what conclusions can you make about the different independent variables in this experiment?

A psychologist interested in auditory psychophysics wants to know if subjects can match pitches of tones presented to different ears as well as they can match pitches of tones presented to the same ear. To investigate this question, she designs the following factorial experiment. She asks subjects to put on headphones and listen as she presents two tones of slightly different pitch. She asks each subject if the second tone is higher or lower in pitch than the first. Her two independent variables are the pitches of the tones--high, medium, or low--and the method of presentation--whether the tones are presented to the same ear or to different ears. Her dependent variable is the number of correct responses the subject makes in identifying the second tone as higher or lower than the first. Use the following data from this experiment to answer problems 23 through 45.

| | Loudness | | | |
|---|---|---|---|---|
| | Soft | Medium | Loud | Row Total |
| Same ear | 12<br>11<br>12 | 17<br>18<br>16 | 21<br>20<br>22 | |
| | $\Sigma X$ =<br>$\Sigma X^2$ = | $\Sigma X$ =<br>$\Sigma X^2$ = | $\Sigma X$ =<br>$\Sigma X^2$ = | |
| Different ears | 9<br>7<br>6 | 13<br>14<br>15 | 16<br>18<br>19 | |
| | $\Sigma X$ =<br>$\Sigma X^2$ = | $\Sigma X$ =<br>$\Sigma X^2$ = | $\Sigma X$ =<br>$\Sigma X^2$ = | |
| | Column Total = | Column Total = | Column Total = | |

$\Sigma\Sigma X$ = $\Sigma\Sigma X^2$ =

23. Compute the sum of squares total.

$SS_{total}$ =

24. Compute the sum of squares within groups.

$SS_{wg}$ =

25. Compute the sum of squares of the rows.

$SS_r$ =

26. Compute the sum of squares of the columns.

$SS_c$ =

27. Compute the sum of squares of the interaction.

$SS_{rXc}$ =

**28.** Compute degrees of freedom total.

$df_{\text{total}} =$

**29.** Compute the degrees of freedom within groups.

$df_{\text{wg}} =$

**30.** Compute the degrees of freedom of the rows.

$df_{\text{r}} =$

**31.** Compute the degrees of freedom of the columns.

$df_{\text{c}} =$

**32.** Compute the degrees of freedom of the interaction.

$df_{\text{rXc}} =$

33. Compute the mean square within groups.

$MS_{wg}$ =

34. Compute the mean square of the rows.

$MS_r$ =

35. Compute the mean square of the columns.

$MS_c$ =

36. Compute the mean square of the interaction.

$MS_{rXc}$ =

37. Compute the $F$ ratio for the rows.

$F_r$ =

**38.** Compute the $F$ ratio for the columns.

$F_{C}$ =

**39.** Compute the $F$ ratio for the interaction.

$F_{rXc}$ =

**40.** Is the $F$ ratio for the rows significant for a *two-tailed* test?

$df$ = ( , )

Critical value =

Significant? Yes No

**41.** Is the $F$ ratio for the columns significant for a *two-tailed* test?

$df$ = ( , )

Critical value =

Significant? Yes No

42. If the $F$ ratio for the columns is significant, then compute an *HSD* so you can determine which of the three loudness conditions are different from one another.

*HSD* =

List the pairs of conditions that are significantly different from one another:

43. Is the $F$ ratio for the interaction significant for a ***two-tailed*** test?

$df$ = ( , )

Critical value =

Significant? Yes No

**44.** Use the space below to graph the means for each cell in order to illustrate the interaction.

**45.** Given the significance of the *F* ratios you just computed, what conclusions can you make about the different independent variables in this experiment?

# CHAPTER 15

# NON-PARAMETRIC STATISTICS

Currently, only medical doctors can prescribe drugs. This includes drugs for the treatment of mental disorders such as schizophrenia and depression. Many clinical psychologists believe that they should also have the legal right to prescribe drugs, when appropriate, for their patients. To see if medical doctors and psychologists are in agreement, a researcher asked 150 medical doctors and 150 clinical psychologists the question, "Should clinical psychologists have the right to prescribe drugs for patients with mental disorders?" The results of this survey follow. Use these data to answer problems 1 through 8.

|  | Should psychologists prescribe drugs? Yes | No | Row Total |
|---|---|---|---|
| Psychologists | A $f_o$=125 $f_e$= | B $f_o$= 25 $f_e$= |  |
| Medical doctors | C $f_o$= 50 $f_e$= | D $f_o$=100 $f_e$= |  |
| Column Total |  |  | Grand total= |

1. Compute the column totals for the survey data and enter them in the table.

2. Compute the row totals for the survey data and enter them in the table.

3. Compute the grand total for the survey data and enter it in the table.

4. Compute the frequency expected for each cell and enter it in the table.

5. Compute $X^2$ for the survey data.

$X^2$ =

6. What are the degrees of freedom for this $X^2$?

   $df$ =

7. Look up the critical value of $X^2$ for this $df$ in Table X and tell if the $X^2$ is significant.

   Value from Table X =

   Significant? Yes No

8. Given the significance of the computed $X^2$, what conclusions can you make about the data presented above?

Parapsychology has been a controversial research area for many years. Many psychologists would agree with the statement that there are no acceptable research data to support the notion of extrasensory abilities. Nevertheless, there is a surprising number of well-educated people who hold fast to their belief in extrasensory powers. A graduate student in psychology wanted to know if a psychology background had much influence on how students feel about ESP. She surveyed graduate students from three different graduate programs to see if they believe in extrasensory perception (ESP). Use her data, which follows, to answer problems 9 through 16.

Believe in ESP?

| Grad Students | Yes | No | Maybe | Row Total |
|---|---|---|---|---|
| Psychology | $f_o$= 11<br>$f_e$= | $f_o$= 25<br>$f_e$= | $f_o$= 5<br>$f_e$= | |
| Sciences | $f_o$= 50<br>$f_e$= | $f_o$= 10<br>$f_e$= | $f_o$= 10<br>$f_e$= | |
| Humanities | $f_o$= 60<br>$f_e$= | $f_o$= 5<br>$f_e$= | $f_o$= 15<br>$f_e$= | |
| Column Total = | | | | Grand Total= |

9. Compute the column totals for the survey data and enter them in the table.

10. Compute the row totals for the survey data and enter them in the table.

11. Compute the grand total for the survey data and enter it in the table.

12. Compute the frequency expected for each cell and enter it in the table.

13. Compute $X^2$ for the data.

    $X^2$ =

14. What are the degrees of freedom for this $X^2$?

    $df$ =

15. Look up the critical value of $X^2$ for the degrees of freedom in Table X and tell if the $X^2$ is significant.

    Value from Table X =

    Significant? Yes No

As a psychologist interested in eating behavior, you decide to determine if scary movies cause people to eat more popcorn than musicals. You randomly assign 10 subjects to watch a scary movie (*Psycho*) and another 9 subjects to watch a musical (*The Sound of Music*). At the beginning of the movie, you give each subject a tub of popcorn. At the end of the movie, you measure the number of pieces of popcorn eaten by each subject. You have decided to analyze the data for this experiment by using the **Mann-Whitney *U***. Use the following data to answer problems 16 through 23.

| Scary $X_1$ | $R_1$ | Musical $X_2$ | $R_2$ |
|---|---|---|---|
| 45 | | 32 | |
| 67 | | 38 | |
| 69 | | 33 | |
| 56 | | 49 | |
| 73 | | 44 | |
| 56 | | 60 | |
| 63 | | 48 | |
| 84 | | 36 | |
| 49 | | 23 | |
| 56 | ____ | | ____ |

$\Sigma R_1 =$ $\Sigma R_2 =$

**16.** Rank the scores and write the ranks in the table above.

**17.** Compute the sum of the ranks for sample 1.

$\Sigma R_1 =$

**18.** Compute the sum of the ranks for sample 2.

$\Sigma R_2$ =

**19.** Compute $U_1$.

$U_1$ =

**20.** Compute $U_2$.

$U_2$ =

**21.** Which of the computed $U$'s, $U_1$ or $U_2$, is $U$?

$U$ =

**22.** Look in Table U to find the critical value of your computed $U$ for a *two-tailed* test. Is your computed $U$ significant?

Table U Value =

Significant? Yes No

**23.** Given the significance of the $U$ statistic you just computed, what are your conclusions about the popcorn-eating data?

A human factors psychologist working for a hammer manufacturer is asked to evaluate the effectiveness of a new hammer design over the traditional hammer design. To test this design, he measures the number of nails missed out of 100 tries for a group of 7 men using the new hammer and compares this to a group of 10 men using a traditional hammer. Use the data that follows and conduct a **Mann-Whitney *U*** to answer problems 24 through 31.

| New | $R_1$ | Traditional | $R_2$ |
|---|---|---|---|
| 84 | | 88 | |
| 80 | | 90 | |
| 88 | | 85 | |
| 91 | | 88 | |
| 90 | | 84 | |
| 83 | | 87 | |
| 83 | | 92 | |
| | | 91 | |
| | | 83 | |
| | | 82 | |
| | $\Sigma R_1=$ | | $\Sigma R_2=$ |

**24.** Rank the scores and write the ranks in the table above.

**25.** Compute the sum of the ranks for sample 1.

$\Sigma R_1 =$

**26.** Compute the sum of the ranks for sample 2.

$\Sigma R_2 =$

**27.** Compute $U_1$.

$U_1$ =

**28.** Compute $U_2$.

$U_2$ =

**29.** Which of the computed $U$'s, $U_1$ or $U_2$, is $U$?

$U$ =

**30.** Look in Table U to find the critical value of your computed $U$ for a *two-tailed* test. Is your computed $U$ significant?

Table U Value =

Significant? Yes No

**31.** Given the significance of the $U$ statistic you just computed, what are your conclusions about the hammer data?

An educational psychologist is interested in the ability of preschool children to solve math story problems. He wants to see if the method of presentation--either as verbal story problems or as visual story problems--makes a difference in preschoolers' abilities to solve the problems correctly. In an example of the verbal condition, a child might be asked, "Two birds are sitting on a fence; two more birds fly down and join them. How many birds are on the fence altogether?" In an example of the nonverbal, visual equivalent of this problem, the experimenter might present the child with a picture of two birds on the fence with two birds in the process of landing on the fence and then ask the child, "How many birds are on the fence altogether?" In both conditions, the the child responds orally. Use the **Wilcoxon *T*** to analyze the following data, which consist of the number of correct problems out of 10, to answer problems 32 through 39.

| Child | Verbal | Nonverbal | *D* | +R | -R |
|---|---|---|---|---|---|
| C. J. | 3 | 6 | | | |
| F. K. | 5 | 8 | | | |
| M. O. | 7 | 9 | | | |
| I. M. | 4 | 8 | | | |
| G. G. | 2 | 4 | | | |
| K. T. | 1 | 1 | | | |
| B. W. | 4 | 3 | | | |
| M. B. | 2 | 8 | | | |
| | | | | $\Sigma$+R= | $\Sigma$-R= |

32. Compute the differences and enter them in the table above.

33. Rank the differences and place the ranks in the appropriate column in the table.

**34.** Sum the plus ranks.

$\Sigma+R$ =

**35.** Sum the minus ranks.

$\Sigma-R$ =

**36.** Which of the sums, the sum of the plus ranks or the sum of the minus ranks, is $T$?

$T$ =

**37.** What is the number of signed ranks?

$n$ =

**38.** Look up the critical value of $T$ in Table W for a *two-tailed* test. Is your computed value of $T$ significant?

Table W Value =

Significant? Yes No

Of all the skills performed by football players, kicking requires the most concentration. A sports psychologist is interested in seeing if self-hypnosis can increase concentration and therefore, improve the kicking distance of a sample of 10 punters. She asks each punter to kick the ball five times and records the farthest of their five kicks. The following day, she teaches the punters a simple self-hypnosis ritual that focuses their concentration on their punting. The next day, the psychologist tells the punters to use self-hypnosis immediately before each of their five kicks. After each punter completes his five kicks, she records the farthest one. Conduct a **Wilcoxon *T*** on the following data to answer problems 39 through 46.

| Punter | Normal | Hypno | $D$ | +R | -R |
|---|---|---|---|---|---|
| C. J. | 45 | 48 | | | |
| F. K. | 40 | 45 | | | |
| M. O. | 33 | 32 | | | |
| I. M. | 35 | 39 | | | |
| G. G. | 47 | 53 | | | |
| K. T. | 52 | 65 | | | |
| B. W. | 47 | 53 | | | |
| M. B. | 46 | 46 | | | |
| K. T. | 39 | 41 | | | |
| B. B. | 28 | 45 | | | |
| | | | | $\Sigma$+R= | $\Sigma$-R= |

**39.** Compute the differences and enter them in the table above.

**40.** Rank the differences and place the ranks in the appropriate column in the table above.

41. Sum the plus ranks.

$\Sigma+R$ =

42. Sum the minus ranks.

$\Sigma-R$ =

43. Which of the sums, the sum of the plus ranks or the sum of the minus ranks, is $T$?

$T$ =

44. What is the number of signed ranks?

$n$ =

45. Look up the critical value of $T$ in Table W for a *two-tailed* test. Is your computed value of $T$ significant?

Table W Value =

Significant? Yes No

46. Given the significance of the computed $T$, what are your conclusions regarding these data?

A group of psychologists is interested in seeing if self-help audio tapes can improve a subject's memory. They randomly assign 8 people to one of three conditions. In the audible condition, subjects listen to a tape that discusses a well-known mnemonic technique at a loudness level that is clearly audible. In the subliminal condition, subjects listen to a tape with the same message, but it is played at such a low volume that none of the subjects can consciously hear the message. In the control condition, subjects "listen," but the tape is blank and has no message. The following day, the subjects return and are presented with a list of words. After a five-minute break, they are given a test to see how many words they remember. The dependent variable in this experiment is the score received on the memory test. Conduct a **Kruskal-Wallis Test** to analyze the following data and answer problems 47 through 52.

Type of Tape

| Audible $X_1$ | $R_1$ | Subliminal $X_2$ | $R_2$ | Control $X_3$ | $R_3$ |
|---|---|---|---|---|---|
| 55 | | 32 | | 30 | |
| 62 | | 30 | | 29 | |
| 49 | | 28 | | 26 | |
| 55 | | 31 | | 33 | |
| 61 | | 33 | | 28 | |
| 58 | | 32 | | 29 | |
| 55 | | 36 | | 33 | |
| $\Sigma R_1=$ | | $\Sigma R_2=$ | | $\Sigma R_3=$ | |

**47.** Combine the samples and rank all the scores; then enter those ranks in the table above.

48. Sum the ranks for samples 1, 2, and 3 and enter those values in the table above.

49. What are the following?

$n_1$ =

$n_2$ =

$n_3$ =

$N_T$ =

50. Compute $H$.

$H$ =

51. Look up the critical value of $H$ in Table X. What degree of freedom did you use? What was the critical value found in the table? Was the $H$ you computed significant?

$df$ =

Critical value from Table X =

Significant? Yes No

52. Given the significance of the $H$ you computed, what conclusions can you draw from these data?

A cognitive psychologist has noticed that complex objects take longer to recognize than simple objects. He believes this difference is due to the fact that recognition of complex objects requires more eye movements than simple objects and therefore, they take more time to process. To test his hypothesis, he monitors the eye movements of four groups of 6 subjects in order to count the number of fixations they make when viewing one of four different shapes: a triangle, a pentagon, an octagon, and a dodecagon (a 12-sided polygon). The number of fixations used by each subject to correctly identify each object is shown in the following table. Conduct a **Kruskal-Wallis Test** to analyze the following data and answer problems 53 through 58.

Shape of Object

| Triangle | | Pentagon | | Octagon | | Dodecagon | |
|---|---|---|---|---|---|---|---|
| $X_1$ | $R_1$ | $X_2$ | $R_2$ | $X_3$ | $R_3$ | $X_4$ | $R_4$ |
| 1 | | 2 | | 2 | | 5 | |
| 2 | | 3 | | 4 | | 6 | |
| 1 | | 4 | | 5 | | 6 | |
| 2 | | 4 | | 5 | | 6 | |
| 2 | | 3 | | 4 | | 5 | |
| 1 | | 3 | | 5 | | 7 | |
| $\Sigma R_1=$ | | $\Sigma R_2=$ | | $\Sigma R_3=$ | | $\Sigma R_4=$ | |

**53.** Combine the samples and rank all the scores; then enter those ranks in the preceding table.

**54.** Sum the ranks for samples 1, 2, and 3 and enter those values in the preceding table.

**55.** What are the following?

$n_1$ =

$n_2$ =

$n_3$ =

$n_4$ =

$N_{total}$ =

**56.** Compute *H*.

*H* =

**57.** Look up the critical value of *H* in Table X. What degrees of freedom did you use? What was the critical value found in the table? Was the *H* you computed significant?

*df* =

Critical value from Table X =

Significant? Yes No

**58.** Given the significance of the *H* you computed, what conclusions can you draw from this data?

# APPENDIX

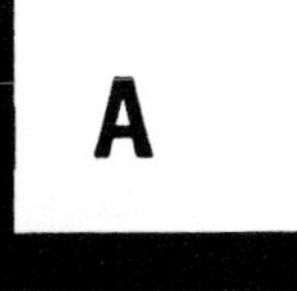


# SOLUTIONS TO PROBLEMS

## CHAPTER 0

1. $<$
2. $Y$ is about equal to $X$
3. 6/2
4. the absolute value of −23
   23
5. a • a
6. 20
7. 23
8. 40
9. 10
10. −40
11. −10
12. 30
13. 0

14. 18

15. 91

16. No. $xy^2$ means $x$ times $y^2$, whereas $(xy)^2$ means $xy$ times $xy$.

17. No. $(x + y)^2 = (x + y)$ times $(x + y) = x^2 + 2xy + y^2$

18. $4\sqrt{x}$

19. 7

20. 6

21. 6

22. 41

23. $\frac{9}{4}$

24. $\frac{5}{4}$

25. $\frac{99}{50}$

# CHAPTER 1

1. 234.255
2. 1.767
3. 8.898
4. 4567.987
5. 777.984
6. 29.889
7. 100.000
8. 1234.568
9. 987.955
10. 0.000
11. Research hypothesis: Adults exhibit a different number of personality traits than do 3-year-olds.

    Null hypothesis: There is no difference between adults and 3-year-olds in the number of personality traits they exhibit.

12. Research hypothesis: Mind-altering drugs can enhance human performance.

    Null hypothesis: Human performance is not enhanced by mind-altering drugs.

13. Research hypothesis: Women of one culture are more aggressive than women of another culture.

    Null hypothesis: There is no difference in the aggressiveness of women from two separate cultures.

14. Research hypothesis: Stress causes a change in the level of a specific nervous system hormone.

    Null hypothesis: Stress has no effect on the level of a specific nervous system hormone.

15. Research hypothesis: People behave differently when they know they are being observed.

    Null hypothesis: There is no change in people's behavior when they know they are being observed.

16. Research Hypothesis: Method A is better for teaching preschoolers than Method B.

    Null hypothesis: There is no difference between Method A and Method B for teaching preschoolers.

17. Independent variable: The type of list (numbers or words)

    Dependent variable: The number of items remembered

18. Independent variable: The size of the group (small, large)

    Dependent variable: Amount of conversation

19. Independent variable: Type of relaxation technique

    Dependent variable: Change in blood pressure

20. Independent variable: Level of hormone

    Dependent variable: Mating behavior

21. Independent variable: Type of problem-solving technique (new or standard)

    Dependent variable: Time to solve problem

22. Nominal

23. Ratio

24. Interval

25. Ratio

26. Ratio

27. Ordinal

28. Ratio

29. Ratio

30. Ratio

31. Ratio

# CHAPTER 2

1. 567 444 438 358 283 236 234 234 129 123

2. 98 92 78 76 67 67 56 43 36 25 24 23 22 10

3. 98 98 67 66 56 44 42 32 30 29 23 10 10

4. 236 123 98 80 67 67 65 45 34 23 22 15 10 8

5. 23 13 10 9 9 9 8 7 6 6 5 5 5 4 4 3 3 2

6. 3

7. 9

8. 45

9. 5

10. 50

11. Number of intervals = 9.388 or 10

    Acceptable? Yes

12. Number of intervals = 10.013

    Acceptable? Yes

13. Number of intervals = 21.457

    Acceptable? No

14. Number of intervals = 15.02

    Acceptable? Yes

15. Number of intervals = 16.689

Acceptable? Yes

16.

| *Real Lim* | *App Lim* | *F* | *RF* | *CF* | *CRF* |
|---|---|---|---|---|---|
| 559.5-599.5 | 560-599 | 2 | .080 | 25 | 1.000 |
| 519.5-559.5 | 520-559 | 1 | .040 | 23 | .920 |
| 479.5-519.5 | 480-519 | 1 | .040 | 22 | .880 |
| 439.5-479.5 | 440-479 | 4 | .160 | 21 | .840 |
| 399.5-439.5 | 400-439 | 2 | .080 | 17 | .680 |
| 359.5-399.5 | 360-339 | 1 | .040 | 15 | .600 |
| 319.5-359.5 | 320-359 | 3 | .120 | 14 | .560 |
| 279.5-319.5 | 280-319 | 2 | .080 | 11 | .440 |
| 239.5-279.5 | 240-279 | 0 | .000 | 9 | .360 |
| 199.5-239.5 | 200-239 | 4 | .160 | 9 | .360 |
| 159.5-199.5 | 160-199 | 1 | .040 | 5 | .200 |
| 119.5-159.5 | 120-159 | 4 | .160 | 4 | .160 |

17.

| *Real Lim* | *App Lim* | *F* | *RF* | *CF* | *CRF* |
|---|---|---|---|---|---|
| 94.5-99.5 | 95-99 | 5 | .100 | 50 | 1.000 |
| 89.5-94.5 | 90-94 | 6 | .120 | 45 | .900 |
| 84.5-89.5 | 85-89 | 5 | .100 | 39 | .780 |
| 79.5-84.5 | 80-84 | 3 | .060 | 34 | .680 |
| 74.5-79.5 | 75-79 | 5 | .100 | 31 | .620 |
| 69.5-74.5 | 70-74 | 9 | .180 | 26 | .520 |
| 64.5-69.5 | 65-69 | 7 | .140 | 17 | .340 |
| 59.5-64.5 | 60-64 | 3 | .060 | 10 | .200 |
| 54.5-59.5 | 55-59 | 4 | .080 | 7 | .140 |
| 49.5-54.5 | 50-54 | 1 | .020 | 3 | .060 |
| 44.5-49.5 | 45-49 | 1 | .020 | 2 | .040 |
| 39.5-44.5 | 40-44 | 1 | .020 | 1 | .020 |

18.

| *Real Lim* | *App Lim* | *F* | *RF* | *CF* | *CRF* |
|---|---|---|---|---|---|
| 95.5-101.5 | 96-101 | 1 | .025 | 40 | 1.000 |
| 89.5- 95.5 | 90- 95 | 1 | .025 | 39 | .975 |
| 83.5- 89.5 | 84- 89 | 5 | .125 | 38 | .950 |
| 77.5- 83.5 | 78- 83 | 7 | .175 | 33 | .825 |
| 71.5- 77.5 | 72- 77 | 2 | .050 | 26 | .650 |
| 65.5- 71.5 | 66- 71 | 2 | .050 | 24 | .600 |
| 59.5- 65.5 | 60- 65 | 2 | .050 | 22 | .550 |
| 53.5- 59.5 | 54- 59 | 3 | .075 | 20 | .500 |
| 47.5- 53.5 | 48- 53 | 3 | .075 | 17 | .425 |
| 41.5- 47.5 | 42- 47 | 2 | .050 | 14 | .350 |
| 35.5- 41.5 | 36- 41 | 2 | .050 | 12 | .300 |
| 29.5- 35.5 | 30- 35 | 2 | .050 | 10 | .250 |
| 23.5- 29.5 | 24- 29 | 2 | .050 | 8 | .200 |
| 17.5- 23.5 | 18- 23 | 3 | .075 | 6 | .150 |
| 11.5- 17.5 | 12- 17 | 3 | .075 | 3 | .075 |

19.

| *Real Lim* | *App Lim* | *F* | *RF* | *CF* | *CRF* |
|---|---|---|---|---|---|
| 83.5-90.5 | 84-90 | 3 | .086 | 35 | 1.000 |
| 76.5-83.5 | 77-83 | 3 | .086 | 32 | .914 |
| 69.5-76.5 | 70-76 | 3 | .086 | 29 | .829 |
| 62.5-69.5 | 63-69 | 3 | .086 | 26 | .743 |
| 55.5-62.5 | 56-62 | 3 | .086 | 23 | .657 |
| 48.5-55.5 | 49-55 | 5 | .143 | 20 | .571 |
| 41.5-48.5 | 42-48 | 2 | .057 | 15 | .429 |
| 34.5-41.5 | 35-41 | 3 | .086 | 13 | .371 |
| 27.5-34.5 | 28-34 | 0 | .000 | 10 | .286 |
| 20.5-27.5 | 21-27 | 3 | .086 | 10 | .286 |
| 13.5-20.5 | 14-20 | 6 | .171 | 7 | .200 |
| 6.5-13.5 | 7-13 | 1 | .029 | 1 | .029 |

20.

| *Real Lim* | *App Lim* | *F* | *RF* | *CF* | *CRF* |
|---|---|---|---|---|---|
| 25.5-27.5 | 26-27 | 1 | .017 | 60 | 1.000 |
| 23.5-25.5 | 24-25 | 2 | .033 | 59 | .983 |
| 21.5-23.5 | 22-23 | 5 | .083 | 57 | .950 |
| 19.5-21.5 | 20-21 | 5 | .083 | 52 | .867 |
| 17.5-19.5 | 18-19 | 9 | .150 | 47 | .783 |
| 15.5-17.5 | 16-17 | 2 | .033 | 38 | .633 |
| 13.5-15.5 | 14-15 | 5 | .083 | 36 | .600 |
| 11.5-13.5 | 12-13 | 7 | .117 | 31 | .517 |
| 9.5-11.5 | 10-11 | 5 | .083 | 24 | .400 |
| 7.5-9.5 | 8-9 | 4 | .067 | 19 | .317 |
| 5.5-7.5 | 6-7 | 5 | .083 | 15 | .250 |
| 3.5-5.5 | 4-5 | 4 | .067 | 10 | .167 |
| 1.5-3.5 | 2-3 | 5 | .083 | 6 | .100 |
| 0-1.5 | 0-1 | 1 | .017 | 1 | .017 |

21.

| *Real Lim* | *App Lim* | *F* | *RF* | *CF* | *CRF* |
|---|---|---|---|---|---|
| 25.5-27.5 | 26-27 | 1 | .040 | 25 | 1.000 |
| 23.5-25.5 | 24-25 | 0 | .000 | 24 | .960 |
| 21.5-23.5 | 22-23 | 0 | .000 | 24 | .960 |
| 19.5-21.5 | 20-21 | 0 | .000 | 24 | .960 |
| 17.5-19.5 | 18-19 | 1 | .040 | 24 | .960 |
| 15.5-17.5 | 16-17 | 0 | .000 | 23 | .920 |
| 13.5-15.5 | 14-15 | 0 | .000 | 23 | .920 |
| 11.5-13.5 | 12-13 | 0 | .000 | 23 | .920 |
| 9.5-11.5 | 10-11 | 1 | .040 | 23 | .920 |
| 7.5-9.5 | 8-9 | 2 | .080 | 22 | .880 |
| 5.5-7.5 | 6-7 | 4 | .160 | 20 | .800 |
| 3.5-5.5 | 4-5 | 13 | .640 | 16 | .640 |
| 1.5-3.5 | 2-3 | 3 | .120 | 3 | .120 |

22.

| *Real Lim* | *App Lim* | *F* | *RF* | *CF* | *CRF* |
|---|---|---|---|---|---|
| 944.5-979.5 | 945-979 | 1 | .030 | 33 | 1.000 |
| 909.5-944.5 | 910-944 | 4 | .121 | 32 | .970 |
| 874.5-909.5 | 875-909 | 4 | .121 | 28 | .848 |
| 839.5-874.5 | 840-874 | 6 | .182 | 24 | .727 |
| 804.5-839.5 | 805-839 | 1 | .030 | 18 | .545 |
| 769.5-804.5 | 770-804 | 3 | .091 | 17 | .515 |
| 734.5-769.5 | 735-769 | 1 | .030 | 14 | .424 |
| 699.5-734.5 | 700-734 | 1 | .030 | 13 | .394 |
| 664.5-699.5 | 665-699 | 2 | .061 | 12 | .364 |
| 629.5-664.5 | 630-664 | 2 | .061 | 10 | .303 |
| 594.5-629.5 | 595-629 | 3 | .091 | 8 | .242 |
| 559.5-594.5 | 560-594 | 0 | .000 | 5 | .152 |
| 524.5-559.5 | 525-559 | 1 | .030 | 5 | .152 |
| 489.5-524.5 | 490-524 | 4 | .121 | 4 | .121 |

23.

| *Real Lim* | *App Lim* | *F* | *RF* | *CF* | *CRF* |
|---|---|---|---|---|---|
| 399.5-424.5 | 400-424 | 1 | .020 | 50 | 1.000 |
| 374.5-399.5 | 375-399 | 1 | .020 | 49 | .980 |
| 349.5-374.5 | 350-374 | 4 | .080 | 48 | .960 |
| 324.5-349.5 | 325-349 | 0 | .000 | 44 | .880 |
| 299.5-324.5 | 300-324 | 5 | .100 | 44 | .880 |
| 274.5-299.5 | 275-299 | 4 | .080 | 39 | .780 |
| 249.5-274.5 | 250-274 | 5 | .100 | 35 | .700 |
| 224.5-249.5 | 225-249 | 7 | .140 | 30 | .600 |
| 199.5-224.5 | 200-224 | 10 | .200 | 23 | .460 |
| 174.5-199.5 | 175-199 | 9 | .180 | 13 | .260 |
| 149.5-174.5 | 150-174 | 4 | .080 | 4 | .080 |

24.

| *Real Lim* | *App Lim* | *F* | *RF* | *CF* | *CRF* |
|---|---|---|---|---|---|
| 84.5-89.5 | 85-89 | 3 | .068 | 44 | 1.000 |
| 79.5-84.5 | 80-84 | 3 | .068 | 41 | .932 |
| 74.5-79.5 | 75-79 | 3 | .068 | 38 | .864 |
| 69.5-74.5 | 70-74 | 4 | .091 | 35 | .795 |
| 64.5-69.5 | 65-69 | 2 | .045 | 31 | .705 |
| 59.5-64.5 | 60-64 | 3 | .068 | 29 | .659 |
| 54.5-59.5 | 55-59 | 4 | .091 | 26 | .591 |
| 49.5-54.5 | 50-54 | 2 | .045 | 22 | .500 |
| 44.5-49.5 | 45-49 | 4 | .091 | 20 | .455 |
| 39.5-44.5 | 40-44 | 3 | .068 | 16 | .364 |
| 34.5-39.5 | 35-39 | 4 | .091 | 13 | .295 |
| 29.5-34.5 | 30-34 | 2 | .045 | 9 | .205 |
| 24.5-29.5 | 25-29 | 5 | .114 | 7 | .159 |
| 19.5-24.5 | 20-24 | 2 | .045 | 2 | .059 |

25.

| *Real Lim* | *App Lim* | *F* | *RF* | *CF* | *CRF* |
|---|---|---|---|---|---|
| 35.5-38.5 | 36-38 | 2 | .050 | 40 | 1.000 |
| 32.5-35.5 | 33-35 | 3 | .075 | 38 | .950 |
| 29.5-32.5 | 30-32 | 3 | .075 | 35 | .875 |
| 26.5-29.5 | 27-29 | 2 | .050 | 32 | .800 |
| 23.5-26.5 | 24-26 | 2 | .050 | 30 | .750 |
| 20.5-23.5 | 21-23 | 3 | .075 | 28 | .700 |
| 17.5-20.5 | 18-20 | 4 | .100 | 25 | .625 |
| 14.5-17.5 | 15-17 | 3 | .075 | 21 | .525 |
| 11.5-14.5 | 12-14 | 3 | .075 | 18 | .450 |
| 8.5-11.5 | 9-11 | 9 | .225 | 15 | .375 |
| 5.5-8.5 | 6-8 | 4 | .100 | 6 | .150 |
| 2.5-5.5 | 3-5 | 2 | .050 | 2 | .050 |

# CHAPTER 3

1. *y*, *x*
2. midpoint
3. real limits
4. upper real limit
5.

| *App Lim* | *F* | *CF* | *RF* | *CRF* |
|---|---|---|---|---|
| 99-101 | 2 | 100 | .02 | 1.00 |
| 96-98 | 10 | 98 | .10 | .98 |
| 93-95 | 7 | 88 | .07 | .88 |
| 90-92 | 9 | 81 | .09 | .81 |
| 87-89 | 6 | 72 | .06 | .72 |
| 84-86 | 2 | 66 | .02 | .66 |
| 81-83 | 9 | 64 | .09 | .64 |
| 78-80 | 8 | 55 | .08 | .55 |
| 75-78 | 4 | 47 | .04 | .47 |
| 72-74 | 5 | 43 | .05 | .43 |
| 69-71 | 9 | 38 | .09 | .38 |
| 66-68 | 7 | 29 | .07 | .29 |
| 63-65 | 7 | 22 | .07 | .22 |
| 60-62 | 7 | 15 | .07 | .15 |
| 57-59 | 5 | 8 | .05 | .08 |
| 54-56 | 3 | 3 | .03 | .03 |

6.

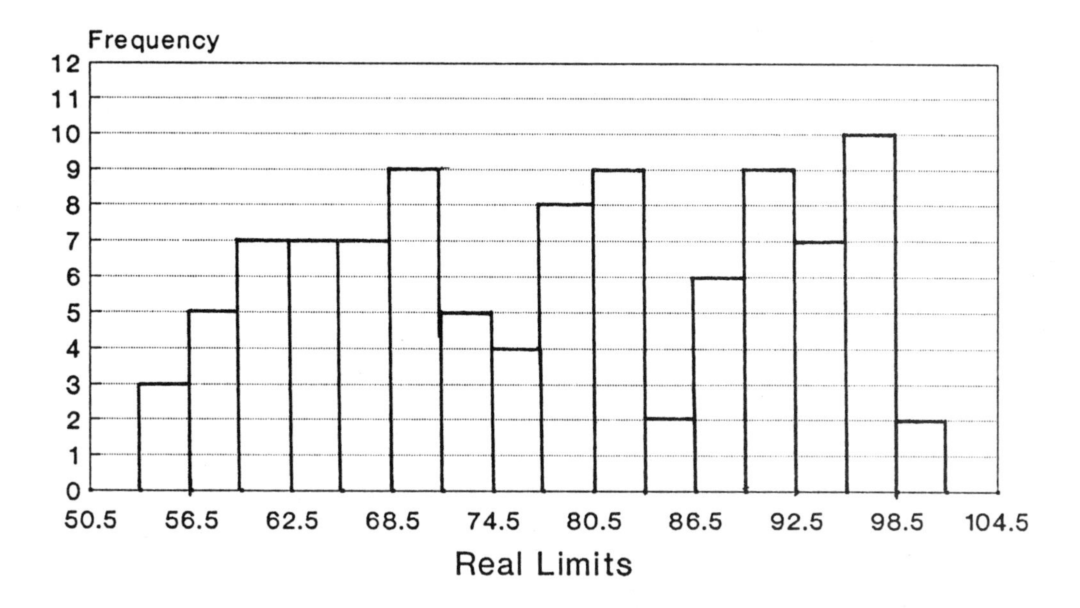
Frequency
12
11
10
9
8
7
6
5
4
3
2
1
0
50.5
56.5
62.5
68.5
74.5
80.5
86.5
92.5
98.5
104.5
Real Limits

7.

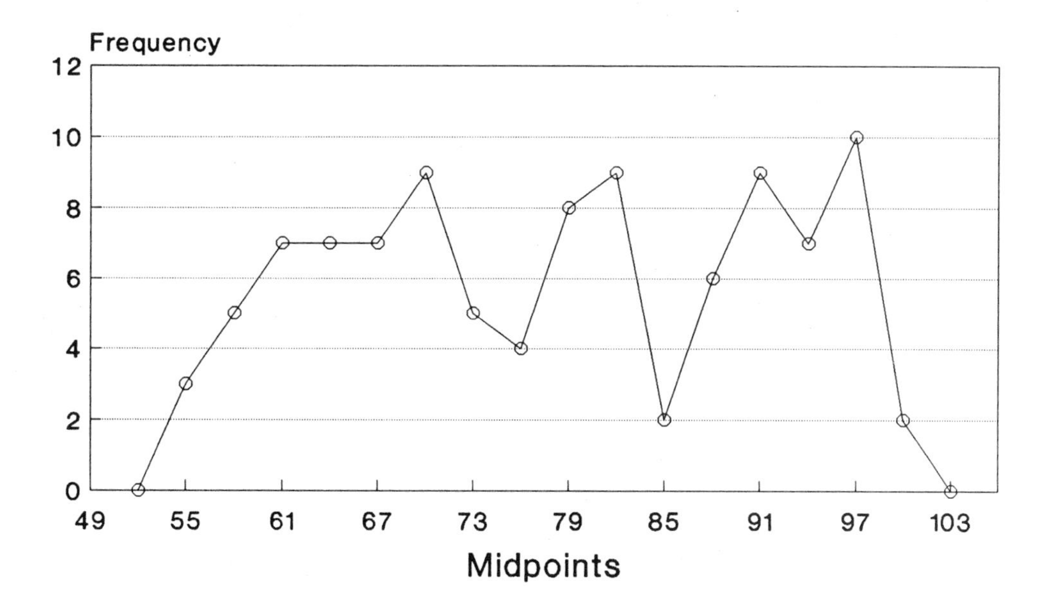
Frequency
12
10
8
6
4
2
0
49
55
61
67
73
79
85
91
97
103
Midpoints

8.

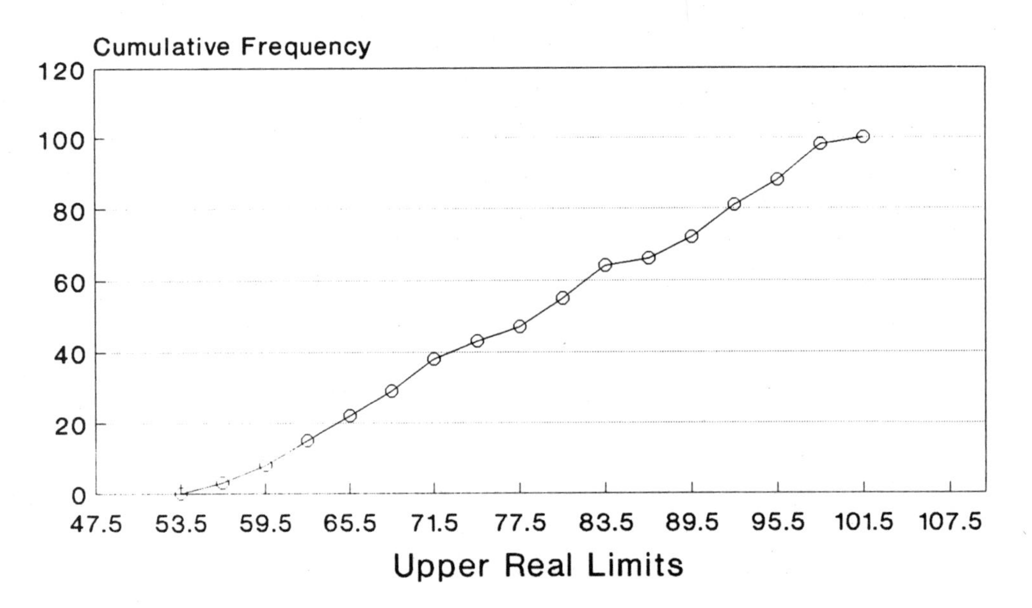

Cumulative Frequency
120
100
80
60
40
20
0
47.5
53.5
59.5
65.5
71.5
77.5
83.5
89.5
95.5
101.5
107.5
Upper Real Limits

9.

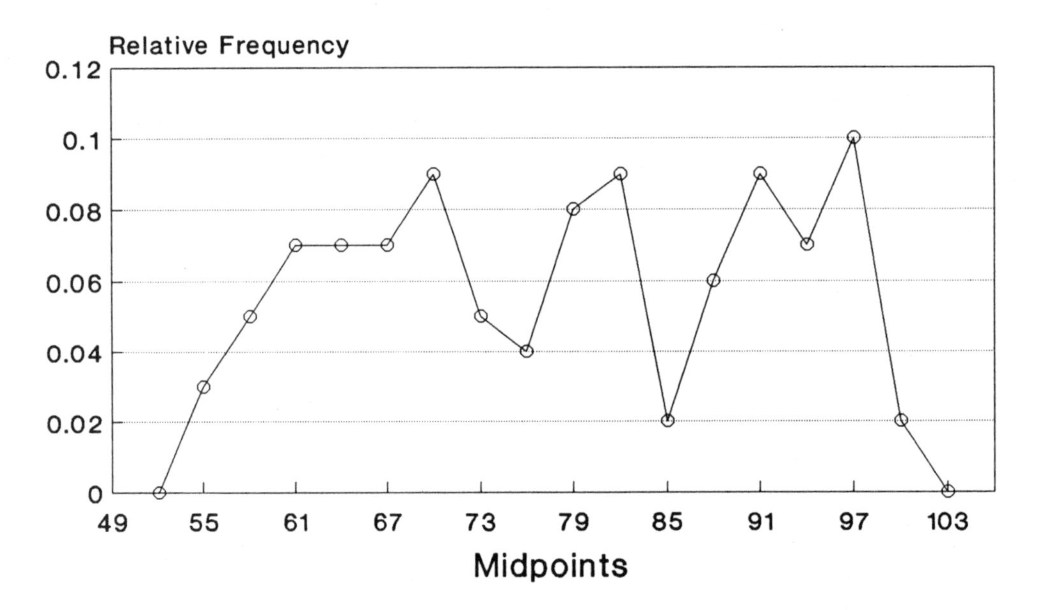

Relative Frequency
0.12
0.1
0.08
0.06
0.04
0.02
0
49
55
61
67
73
79
85
91
97
103
Midpoints

10.

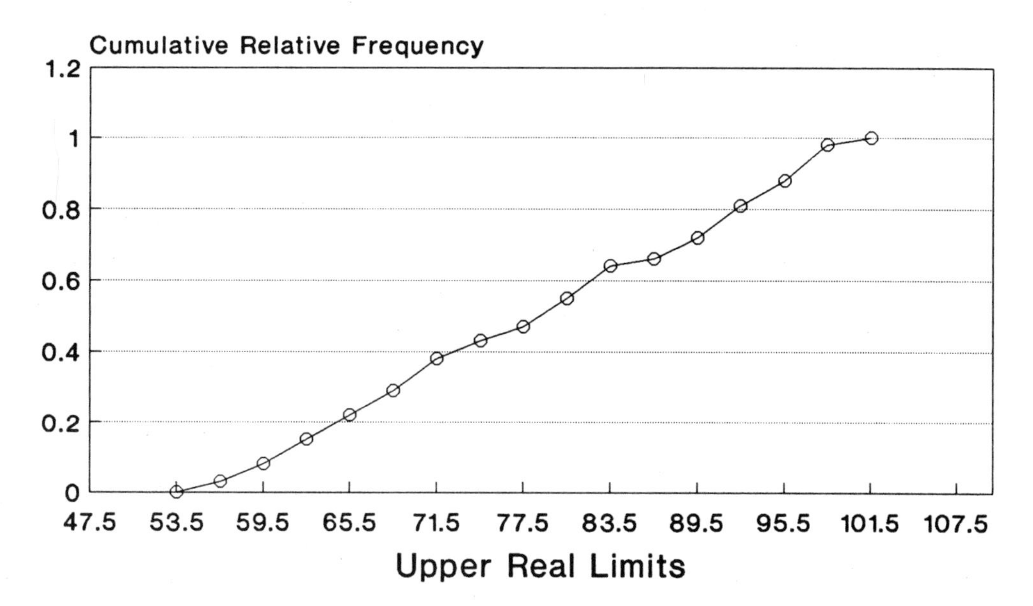

Cumulative Relative Frequency
1.2
1
0.8
0.6
0.4
0.2
0
47.5
53.5
59.5
65.5
71.5
77.5
83.5
89.5
95.5
101.5
107.5
Upper Real Limits


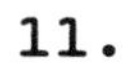

11.

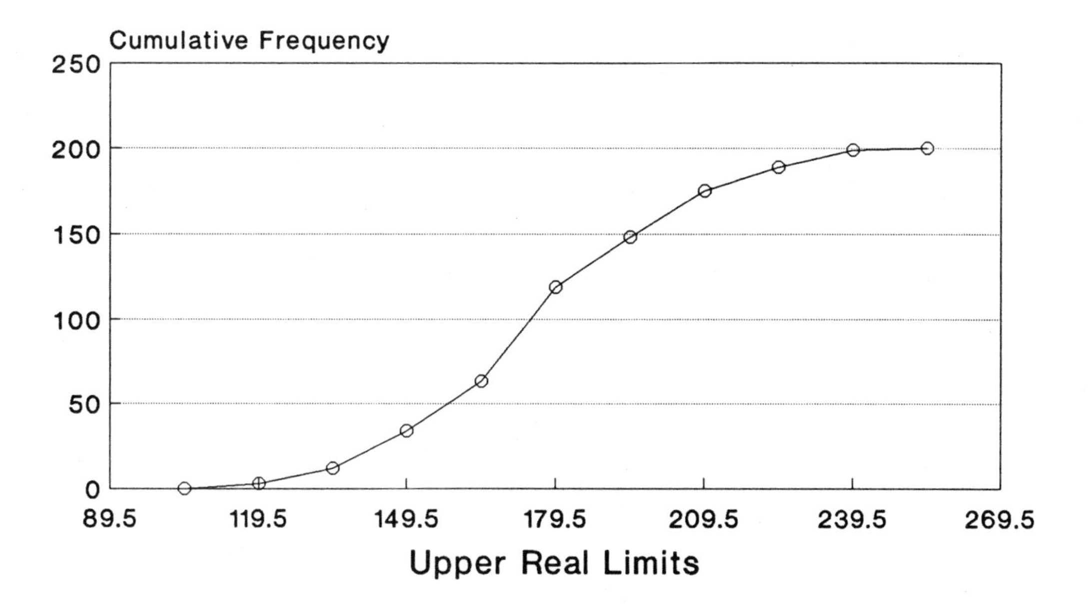

Cumulative Frequency
250
200
150
100
50
0
89.5
119.5
149.5
179.5
209.5
239.5
269.5
Upper Real Limits

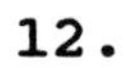

12.

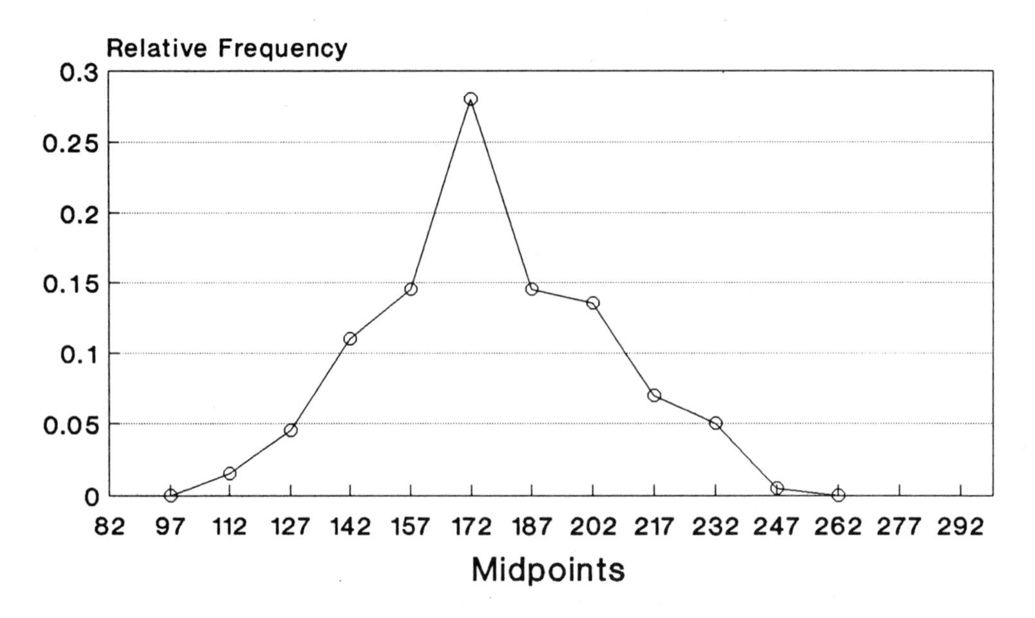

Relative Frequency
0.3
0.25
0.2
0.15
0.1
0.05
0
82 97 112 127 142 157 172 187 202 217 232 247 262 277 292
Midpoints


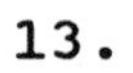

13.

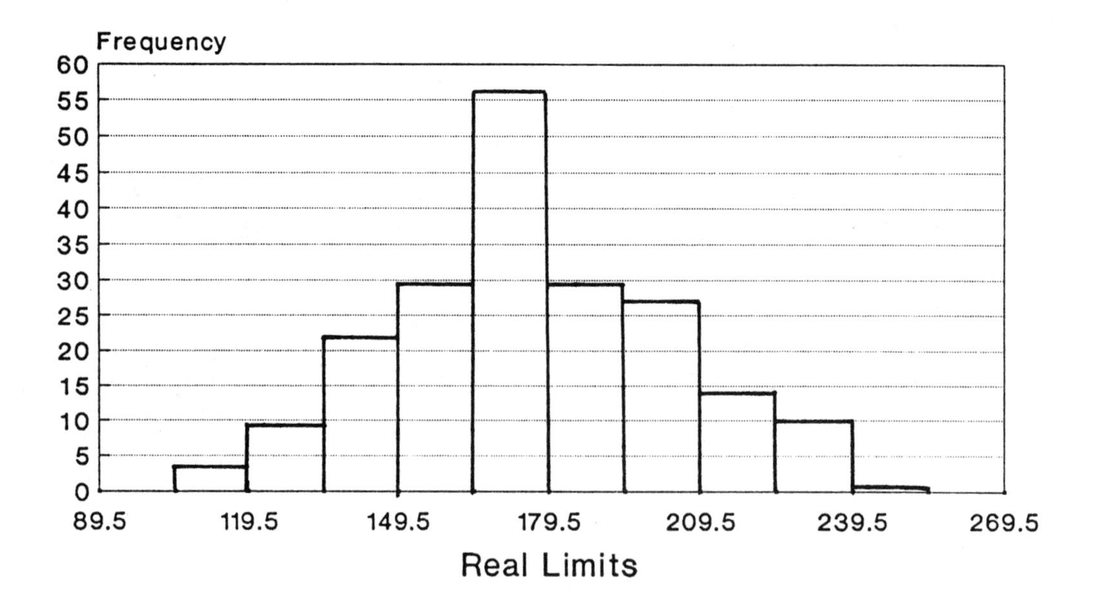

Frequency
60
55
50
45
40
35
30
25
20
15
10
5
0
89.5 119.5 149.5 179.5 209.5 239.5 269.5
Real Limits

14.

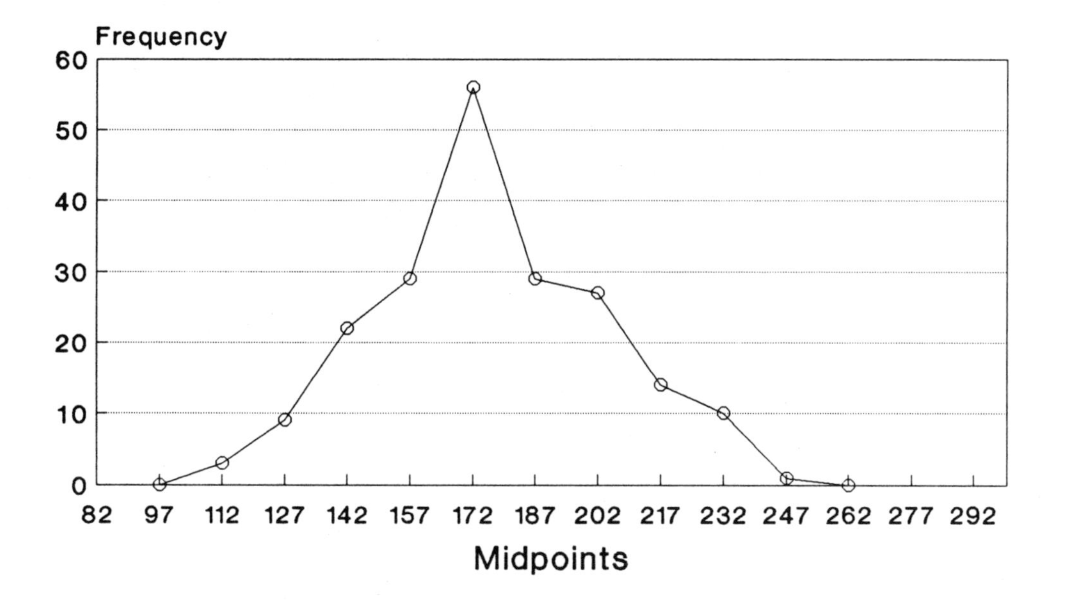
Frequency
60
50
40
30
20
10
0
82 97 112 127 142 157 172 187 202 217 232 247 262 277 292
Midpoints

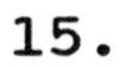

15.

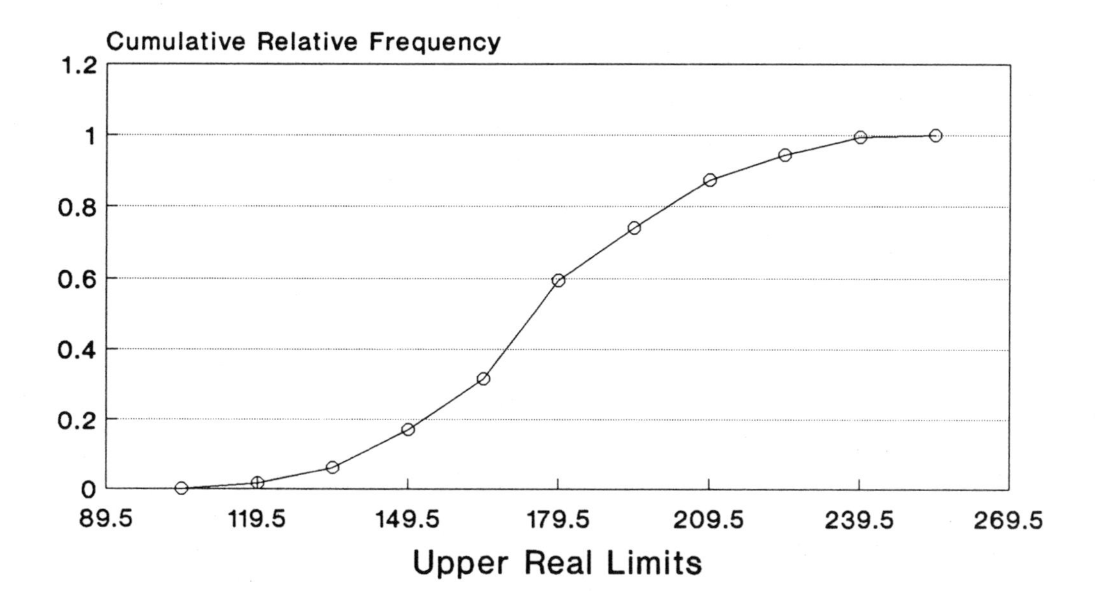
Cumulative Relative Frequency
1.2
1
0.8
0.6
0.4
0.2
0
89.5 119.5 149.5 179.5 209.5 239.5 269.5
Upper Real Limits

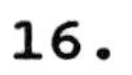

16.

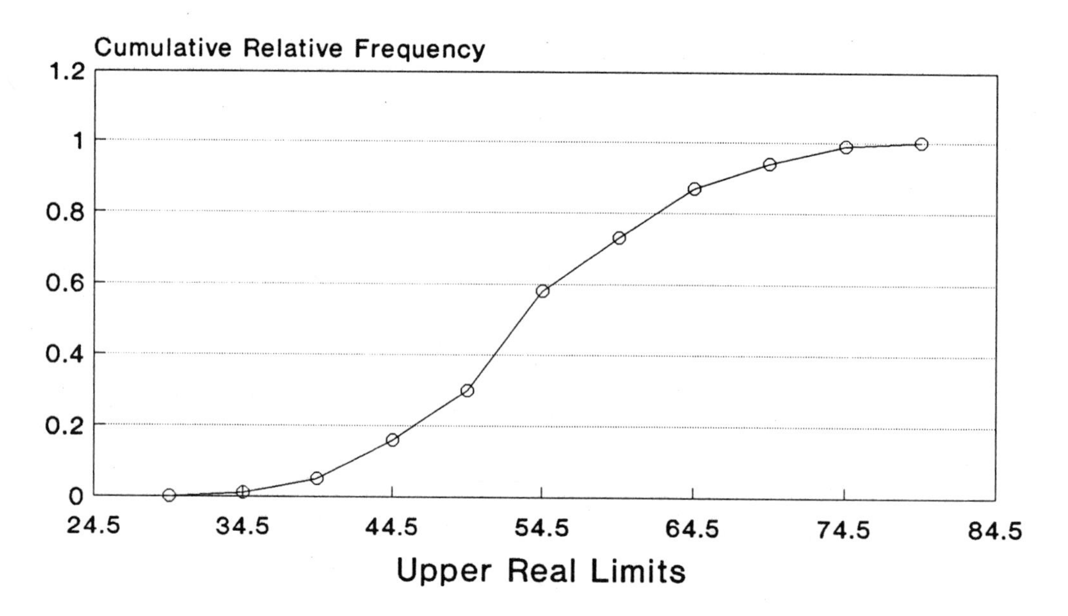

Cumulative Relative Frequency
1.2
1
0.8
0.6
0.4
0.2
0
24.5
34.5
44.5
54.5
64.5
74.5
84.5
Upper Real Limits


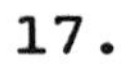

17.

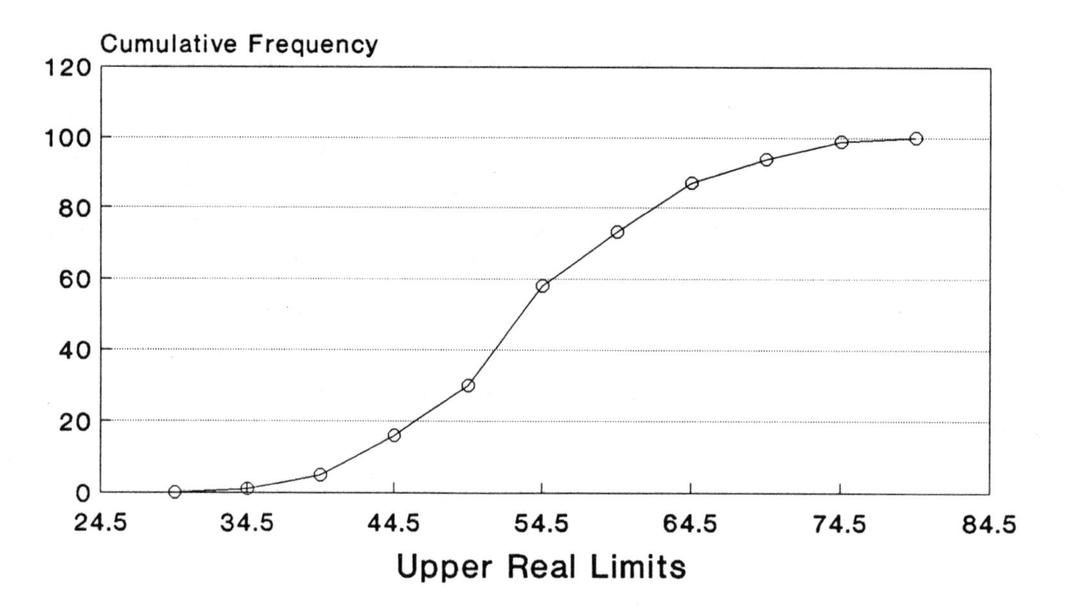

Cumulative Frequency
120
100
80
60
40
20
0
24.5
34.5
44.5
54.5
64.5
74.5
84.5
Upper Real Limits

18.

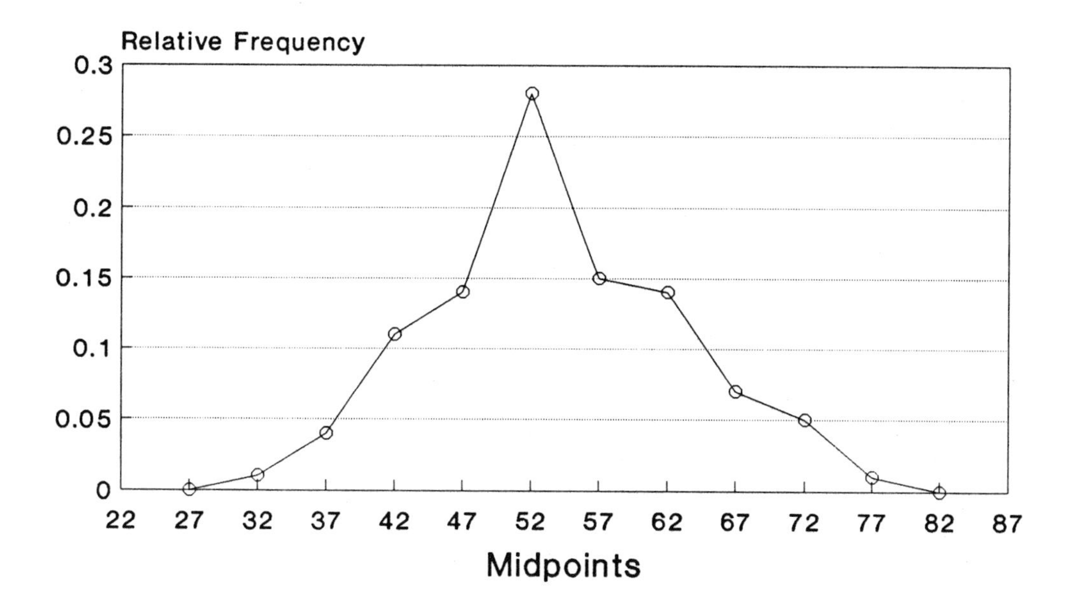
Relative Frequency
0.3
0.25
0.2
0.15
0.1
0.05
0
22 27 32 37 42 47 52 57 62 67 72 77 82 87
Midpoints

19.

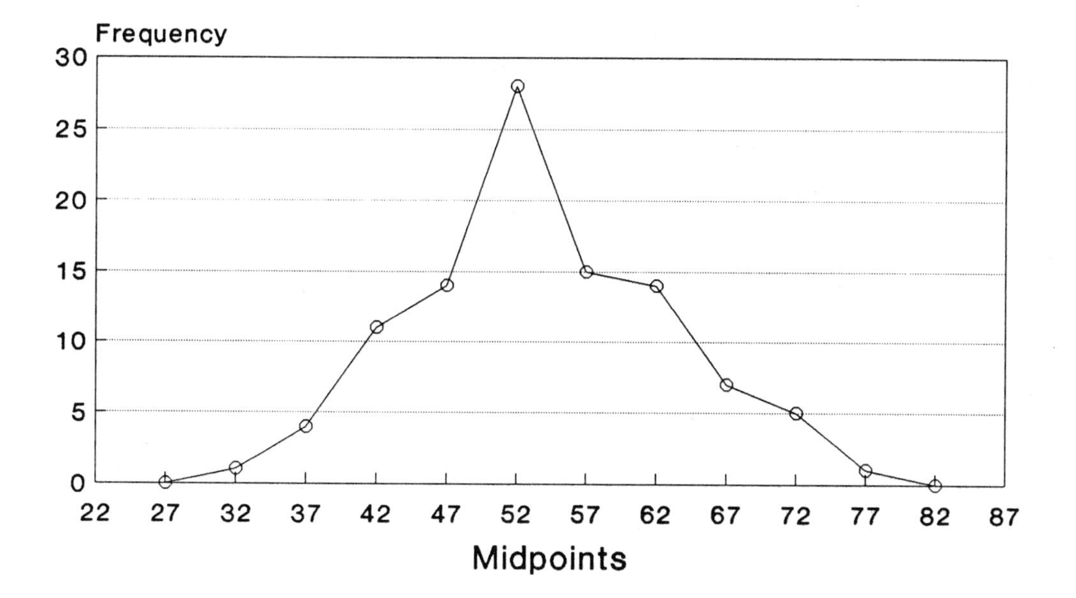
Frequency
30
25
20
15
10
5
0
22 27 32 37 42 47 52 57 62 67 72 77 82 87
Midpoints

20.

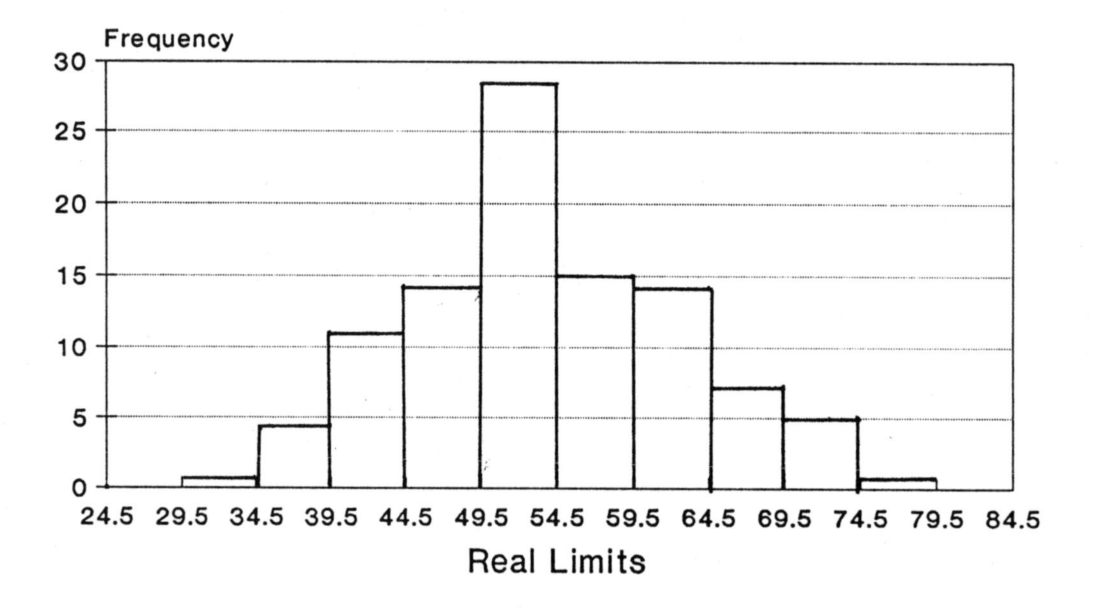
Frequency
30
25
20
15
10
5
0
24.5
29.5
34.5
39.5
44.5
49.5
54.5
59.5
64.5
69.5
74.5
79.5
84.5
Real Limits

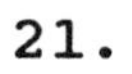
21.

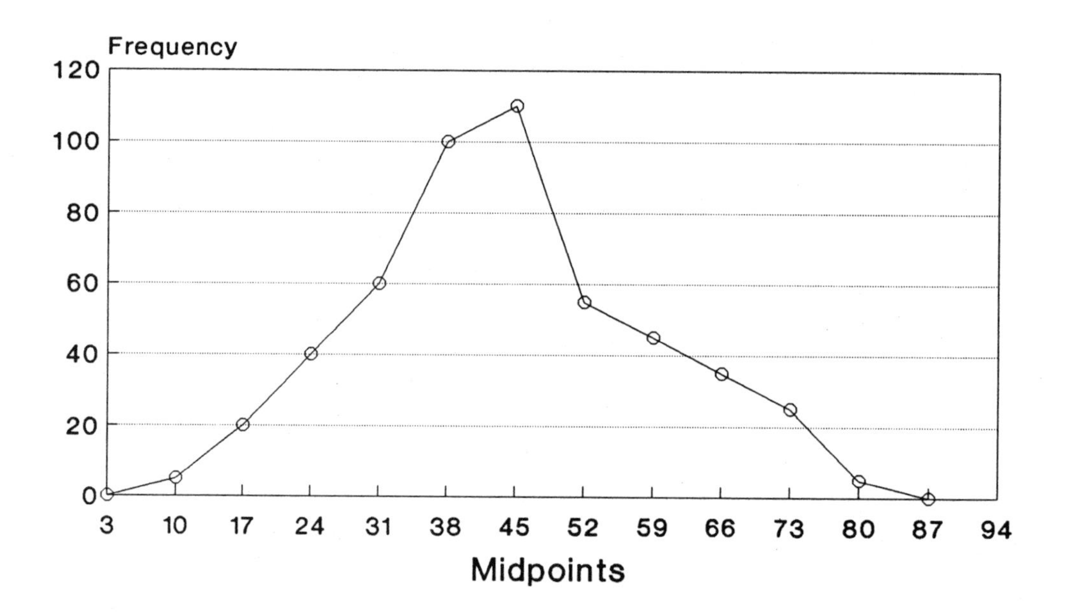
Frequency
120
100
80
60
40
20
0
3
10
17
24
31
38
45
52
59
66
73
80
87
94
Midpoints

22.

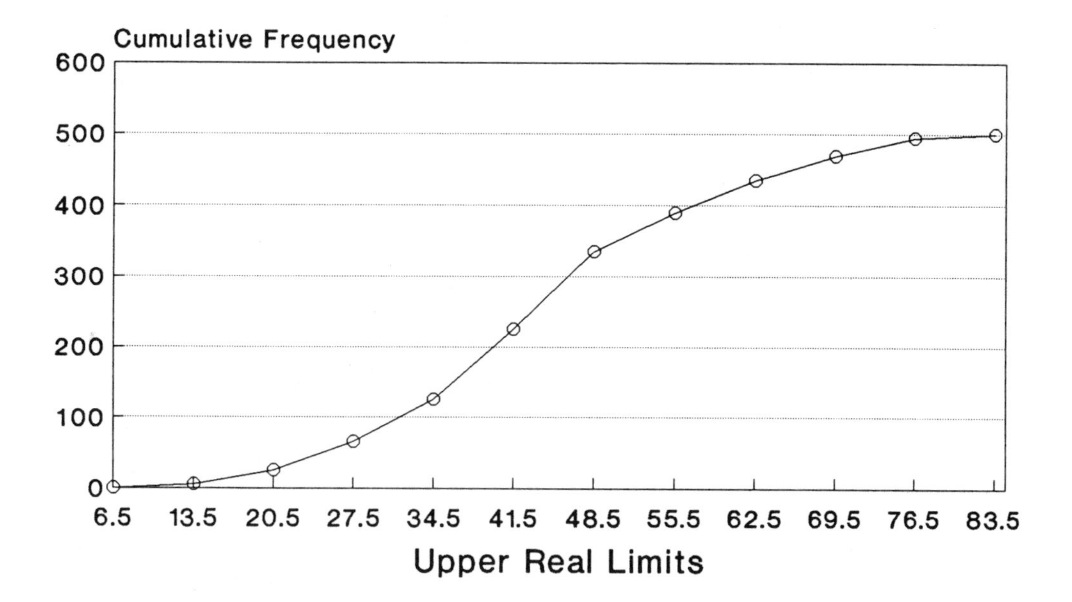
Cumulative Frequency
600
500
400
300
200
100
0
6.5
13.5
20.5
27.5
34.5
41.5
48.5
55.5
62.5
69.5
76.5
83.5
Upper Real Limits

23.

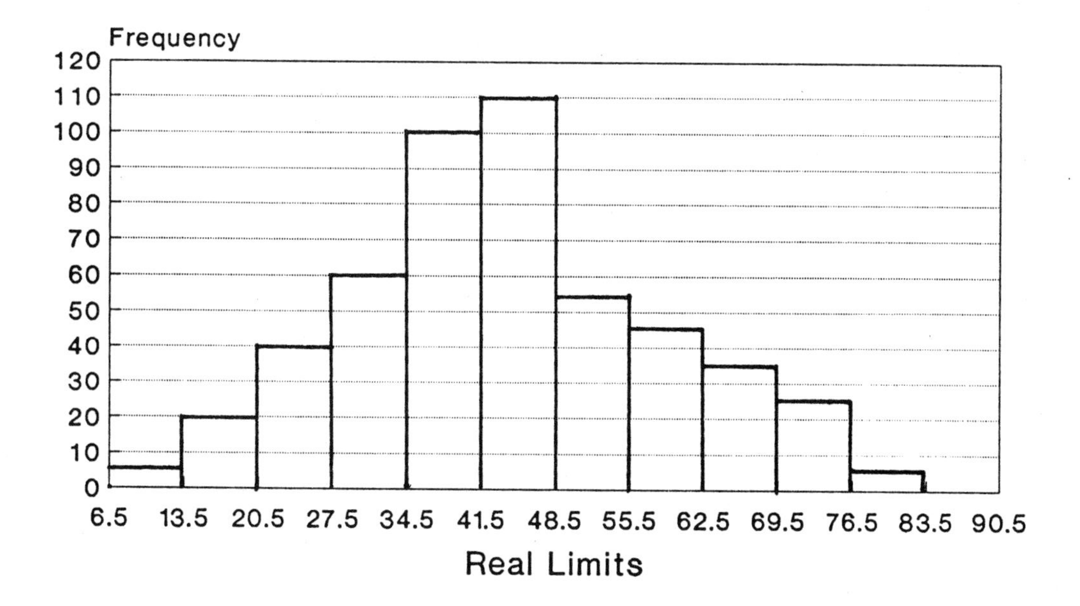
Frequency
120
110
100
90
80
70
60
50
40
30
20
10
0
6.5
13.5
20.5
27.5
34.5
41.5
48.5
55.5
62.5
69.5
76.5
83.5
90.5
Real Limits

24.

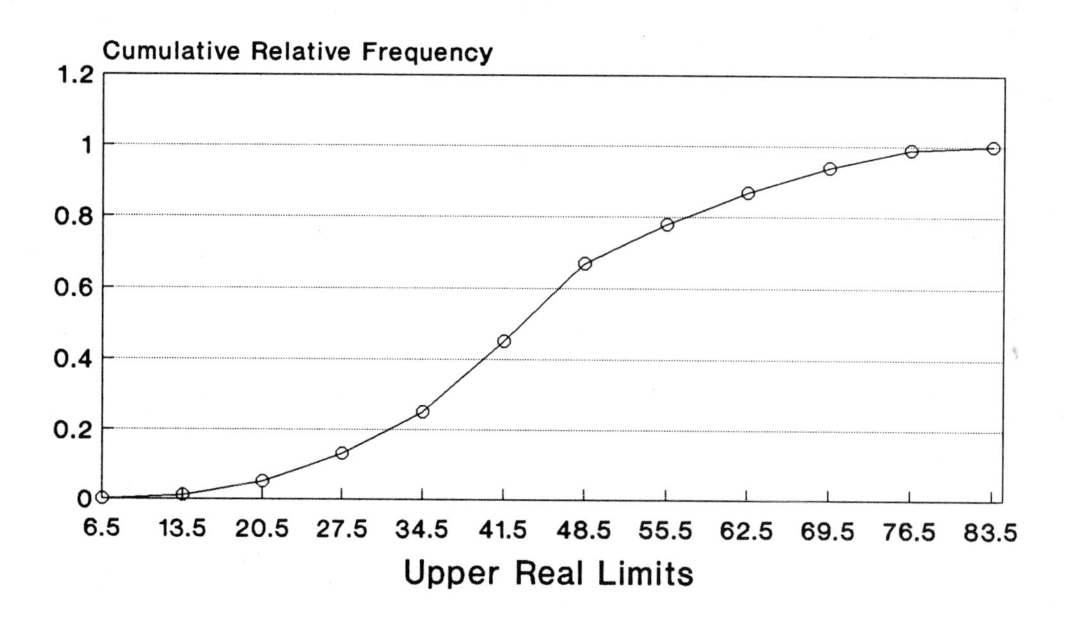
Cumulative Relative Frequency
1.2
1
0.8
0.6
0.4
0.2
0
6.5
13.5
20.5
27.5
34.5
41.5
48.5
55.5
62.5
69.5
76.5
83.5
Upper Real Limits

25.

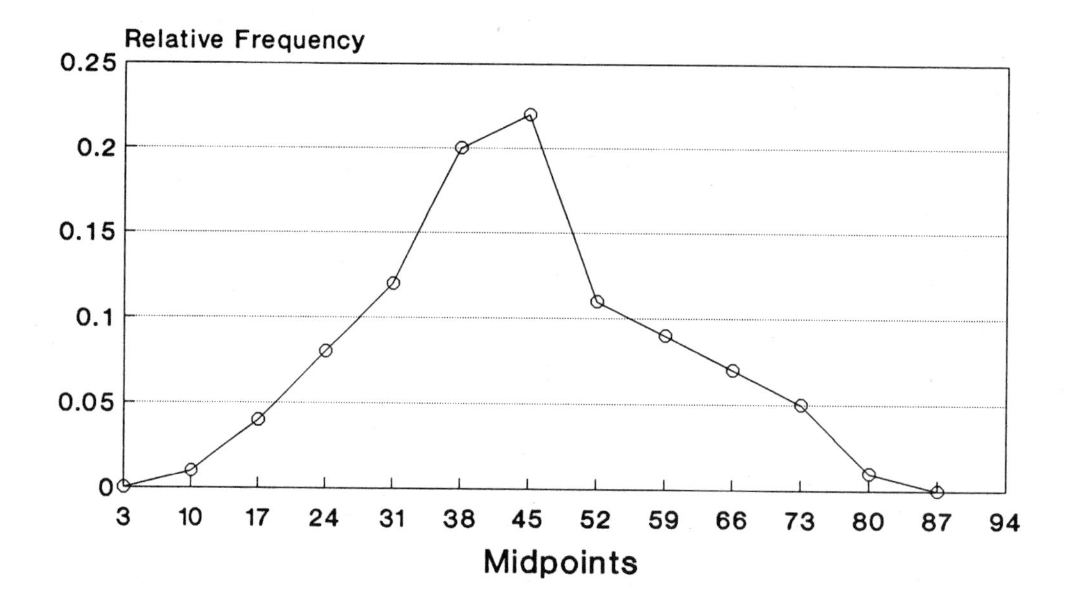
Relative Frequency
0.25
0.2
0.15
0.1
0.05
0
3
10
17
24
31
38
45
52
59
66
73
80
87
94
Midpoints

## CHAPTER 4

1. 88.818
2. 89.5
3. 98
4. negative
5. 35.9
6. 37
7. 59
8. negative
9. 25.733
10. 22
11. 19
12. positive
13. 8
14. 8
15. 12
16. negative
17. 6.44
18. 5
19. 4
20. positive
21. 176.35
22. 174.411
23. 172
24. positive
25. 56.25
26. 53.071
27. 52
28. positive

# CHAPTER 5

1. 326.454
2. 16002.430
3. 126.501
4. 91.412
5. 21.536
6. 4.641
7. 24.9
8. 168.89
9. 12.966
10. 25.733
11. 114.196
12. 10.686
13. 6.462
14. 18.095
15. 4.254
16. 9.733
17. 55.796
18. 7.470
19. 556.375
20. 2555.734
21. 50.554
22. 251.917
23. 7373.743
24. 85.871
25. 176.35
26. 761.828
27. 27.601
28. Company mail produced the most consistent results, because the standard deviation was smaller.
29. The company mail, because it produced the highest mean response rate and it was the most consistent.
30. Electronic mail is the most likely to have the single highest response rate because of its high standard deviation.
31. Again, electronic mail is most likely to have the single lowest response response because of its high standard deviation.

# CHAPTER 6

1. 82, 8
2. 62, 8
3. 77, 8
4. 64.5, 8
5. 144, 16
6. 18, 2
7. 360, 40
8. 216, 24
9. 0, 8
10. 6, 1
11. 0, 1
12. 326.455
13. 126.501
14.

| $X$ | $z$ |
|---|---|
| 223 | -0.818 |
| 234 | -0.731 |
| 234 | -0.731 |
| 567 | 1.902 |
| 236 | -0.715 |
| 129 | -1.561 |
| 438 | 0.882 |
| 283 | -0.344 |
| 444 | 0.929 |
| 358 | 0.249 |
| 445 | 0.937 |

15. 91.412

16. 4.641

17.

| $X$ | $z$ |
|---|---|
| 99 | 1.635 |
| 98 | 1.420 |
| 98 | 1.420 |
| 97 | 1.204 |
| 95 | 0.773 |
| 93 | 0.342 |
| 92 | 0.127 |
| 92 | 0.127 |
| 91 | −0.089 |
| 90 | −0.304 |
| 90 | −0.304 |
| 89 | −0.520 |
| 88 | −0.735 |
| 88 | −0.735 |
| 86 | −1.166 |
| 85 | −1.382 |
| 83 | −1.813 |

18. 24.9

19. 12.996

20.

| $X$ | $z$ |
|---|---|
| 12 | −0.993 |
| 14 | −0.839 |
| 17 | −0.608 |
| 18 | −0.531 |
| 19 | −0.454 |
| 23 | −0.146 |
| 25 | 0.008 |
| 28 | 0.239 |
| 34 | 0.700 |
| 59 | 2.624 |

21. 25.733

22. 10.686

23.

| $X$ | $z$ |
|---|---|
| 12 | -1.285 |
| 14 | -1.098 |
| 16 | -0.911 |
| 18 | -0.724 |
| 19 | -0.630 |
| 19 | -0.630 |
| 20 | -0.537 |
| 22 | -0.349 |
| 23 | -0.256 |
| 24 | -0.162 |
| 35 | 0.867 |
| 36 | 0.961 |
| 39 | 1.241 |
| 42 | 1.522 |
| 47 | 1.990 |

24. 6.461

25. 4.253

26.

| $X$ | $z$ |
|---|---|
| 1 | -1.284 |
| 2 | -1.049 |
| 2 | -1.049 |
| 2 | -1.049 |
| 2 | -1.049 |
| 7 | 0.127 |
| 7 | 0.127 |
| 7 | 0.127 |
| 8 | 0.362 |
| 9 | 0.597 |
| 10 | 0.832 |
| 12 | 1.302 |
| 15 | 2.007 |

27. 9.733

28. 7.470

29.

| $X$ | $z$ |
|---|---|
| 2 | -1.035 |
| 2 | -1.035 |
| 3 | -0.901 |
| 4 | -0.768 |
| 5 | -0.634 |
| 5 | -0.634 |
| 6 | -0.500 |
| 7 | -0.366 |
| 8 | -0.232 |
| 9 | -0.098 |
| 10 | 0.036 |
| 18 | 1.107 |
| 19 | 1.241 |
| 22 | 1.642 |
| 26 | 2.178 |

# CHAPTER 7

1. .1003
2. .3192
3. .5000
4. .3859
5. .0052
6. .4793
7. .4515
8. 0
9. .3133
10. .4115
11. .0668
12. .5000
13. .7257
14. .6915
15. .5000
16. .1587
17. 72.816 seconds
18. 60 seconds
19. 49.636 seconds
20. .1498
21. .5328
22. .0242
23. .0129
24. .0171
25. .1056
26. .5000
27. .8413
28. .5000
29. .0668
30. .1915
31. .2029
32. .0175
33. .0175
34. 112.256 millimeters
35. 91.202 millimeters
36. 31, because it is the farthest from the mean.
37. 46, because it is the closest to the mean.
38. .50, because half of the scores are above the mean.

# CHAPTER 8

1.

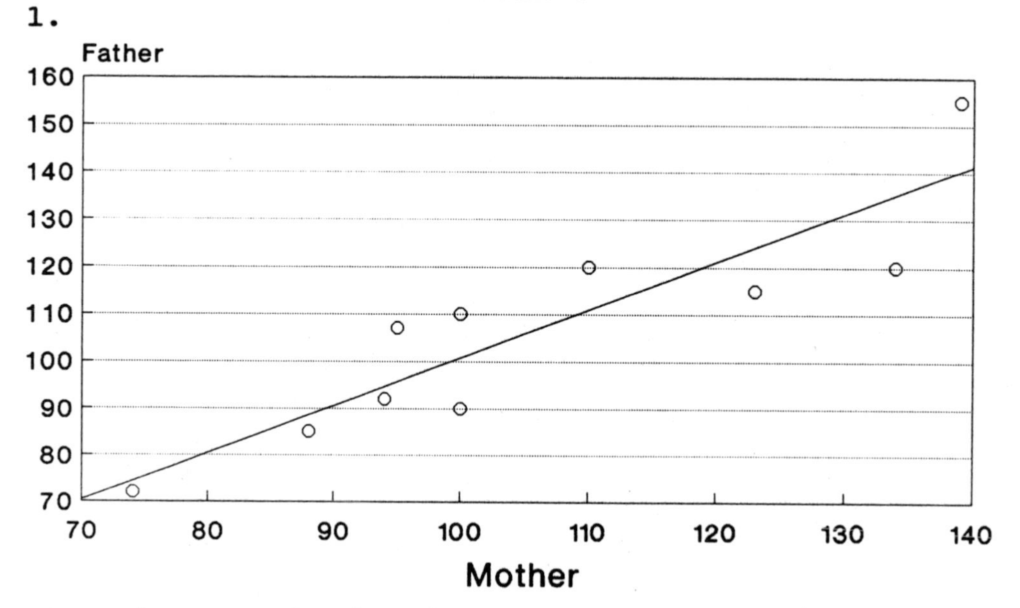


2. The correlation is strong because the scores are bunched relatively tightly along the best fitting straight line.

3. positive

4.

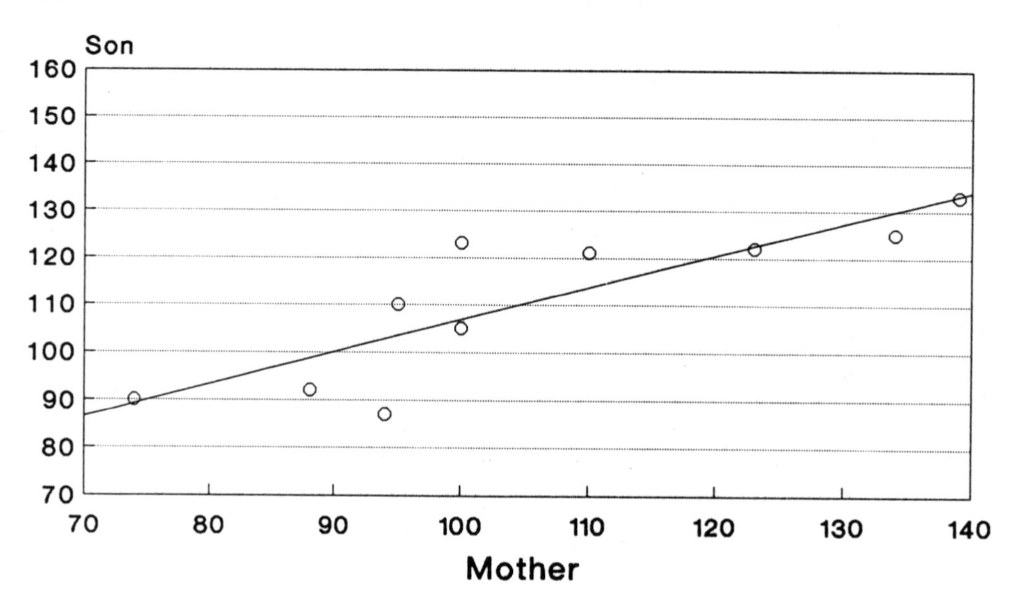

5. The correlation is strong because the scores are bunched relatively tightly along the best fitting straight line.

6. positive

7.

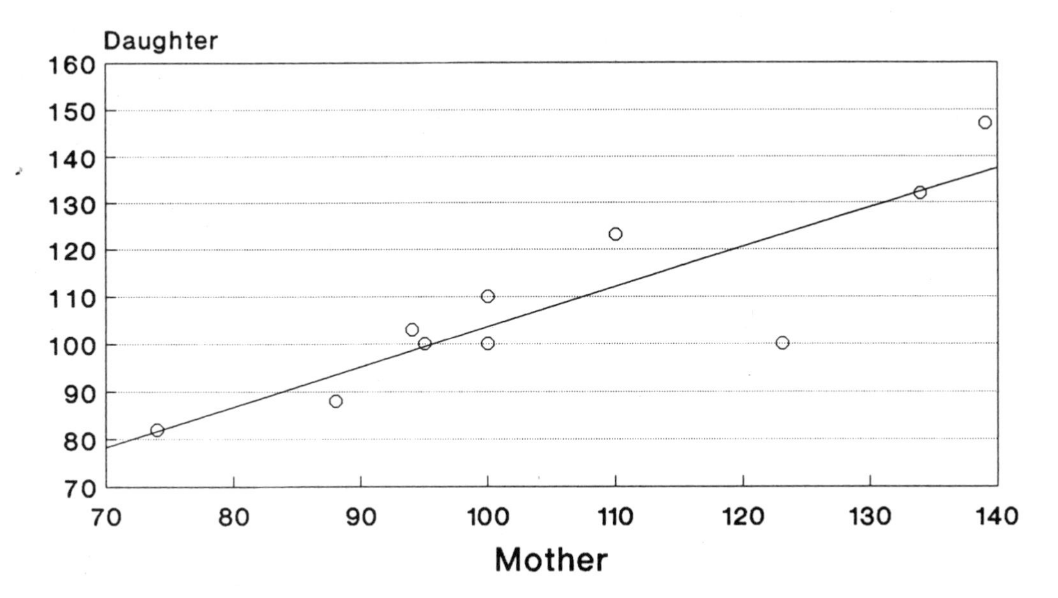


8. The correlation is strong because the scores are bunched relatively tightly along the best fitting straight line.

9. positive

10.

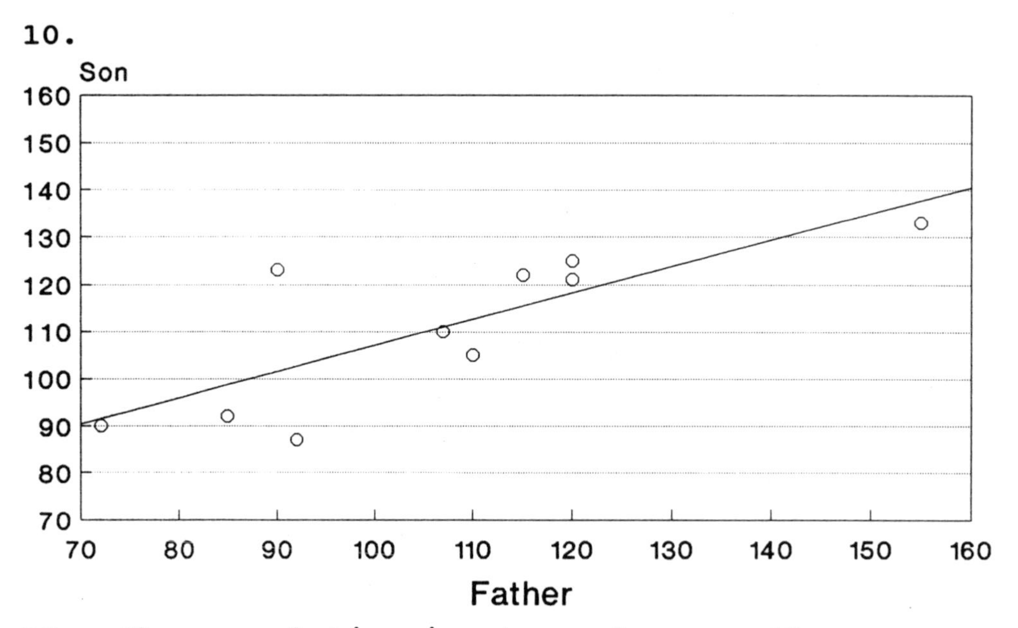


11. The correlation is strong because the scores are bunched relatively tightly along the best fitting straight line.

12. positive

13.

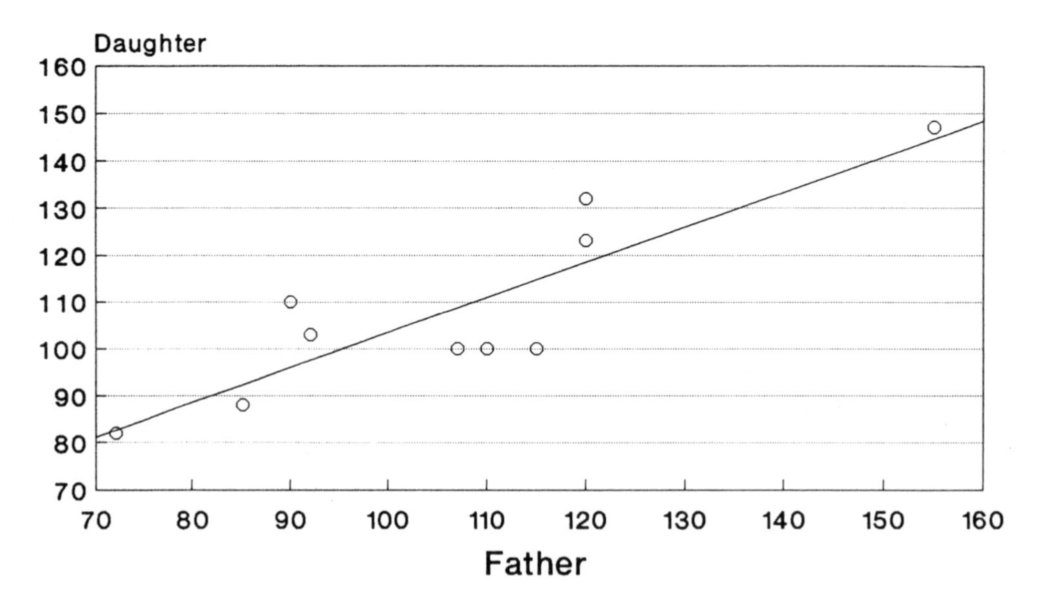

14. The correlation is strong because the scores are bunched relatively tightly along the best fitting straight line.

15. positive

16.

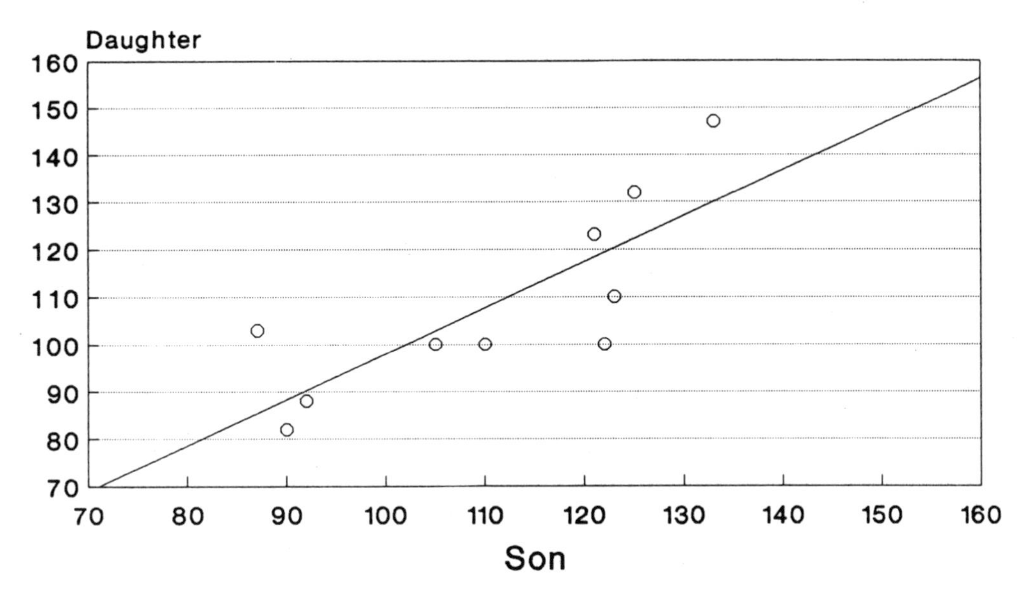


17. The correlation is strong because the scores are bunched relatively tightly along the best fitting straight line.

18. positive

19. $r = .95$

20. $r^2 = .90$ This means that, for these samples, 90% of the variability in IQ scores can be explained by the variability in SAT scores, and vice versa.

21. $r = -.70$

22. $r^2 = .49$ This means that, for these samples, 49% of the variability in GPA can be explained by the variability in birth order, and vice versa.

23. $r = -.64$

24. $r^2 = .41$ This means that, for these samples, 41 % of the variability in birth order can be explained by the variability in IQ scores, and vice versa.

25. $r = -.25$

26. .06 This means that, for these samples, 6% of the variability in IQ can be explained by the variability in shoe size, and vice versa.

27. $r = .92$

28. .85 This means that, for these samples, 85% of the variability in SAT can be explained by the variablilty in GPA, and vice versa.

## CHAPTER 9

1.

$$X' = \left[\frac{.6 \cdot 15}{5}\right] \cdot (Y - 25) + 50$$

$$X' = 1.8\ Y + 5$$

$$X' = 1.8 \cdot 10 + 5 = 18 + 5$$

$$X' = 23$$

2. $X' = 41$

3. $X' = 68$

4.

$$Y' = \left[\frac{.6 \cdot 5}{15}\right] \cdot (X - 50) + 25$$

$$Y' = .2 \cdot X + 15$$

$$Y' = .2 \cdot 60 + 15$$

$$Y' = 27$$

5. $Y' = 28$

6. $Y' = 25$

7.

$$X' = \left[\frac{.75 \cdot 25}{35}\right] \cdot (Y - 125) + 90$$

$$X' = .536 \cdot Y + 23$$

$$X' = .536 \cdot 100 + 23$$

$$X' = 76.6$$

8. $X' = 106.08$

9. $X' = 90$

10.

$$Y' = \left[\frac{.75 \cdot 35}{25}\right] \cdot (X - 90) + 125$$

$$Y' = 1.05 \cdot X + 30.5$$

$$Y' = 1.05 \cdot 120 + 30.5$$

$$Y' = 156.5$$

11. $Y' = 146$

12. $Y' = 125$

13.

$$X' = \left[\frac{-.45 \cdot 20}{5}\right] \cdot (Y - 20) + 200$$

$$X' = -1.84 \cdot Y + 236$$

$$X' = -1.84 \cdot 15 + 236$$

$$X' = 209$$

14. $X' = 227$

15. $X' = 200$

16.

$$Y' = \left[\frac{-.45 \cdot 5}{20}\right] \cdot (X - 200) + 20$$

$$Y' = -.123 \cdot X + 44.6$$

$$Y' = -.123 \cdot 210 + 44.6$$

$$Y' = 18.77$$

17. $Y' = 15.08$

18. $Y' = 20$

19.

$$Y' = \left[\frac{.80 \cdot 6}{5}\right] \cdot (X - 25) + 30$$

$$Y' = .96 \cdot X + 6$$

$$Y' = .06 \cdot 20 + 6$$

$$Y' = 25.2$$

20. $Y' = 20.4$

21. $Y' = 30$

22.

$$X' = \left[\frac{.80 \cdot 5}{6}\right] \cdot (Y - 30) + 25$$

$$X' = .667 \cdot Y + 4.99$$

$$X' = .667 \cdot 20 + 4.99$$

$$X' = 18.33$$

23. $X' = 20.998$

24. $X' = 25$

## CHAPTER 10

1.

$$\text{Probability} = \frac{600}{2000} = .3$$

2. Probability = .125

3. Probability = .01

4. Probability = .25

5. Probability = .025

6. Probability = .1

7. Probability = .05

8. Probability = .05

9. Probability = .225

10. Probability = .425

11. Probability = .45 · .75 = .338

12. Probability = .188

13. Probability = .02

14. Probability = .023

15. Probability = .005

16. Probability = .075

17. Probability = .000 (actually .000225)

18. Probability = .021

19. Probability = .004

20. Probability = .45

21. Probability = .20 + .01 = .21

22. Probability = .1

23. Probability = .03

24. Probability = 1.000

25. Probability = .1

26. Probability = .15

27. Probability = (.55 · .15) + (.45 · .75)

= .083 + .329 = .412

28. Probability = .523

29. Probability = .028

30. Probability = .413

31. $$S^2 = \frac{59350}{5} - 108^2 = 206$$

32. $S = 14.353$

33. $$\text{est. } \sigma^2 = \frac{59350 - (5 \cdot 108^2)}{4 - 1} = 257.5$$

34. est. $\sigma = 16.047$

35. $S^2 = 32.25$

36. $S = 5.679$

37. est. $\sigma^2 = 35.833$

38. est. $\sigma = 5.987$

39. $S^2 = 4.36$

40. $S = 2.088$

41. est. $\sigma^2 = 4.844$

42. est. $\sigma = 2.201$

43. $S^2 = 4.359$

44. $S = 2.088$

45. est. $\sigma^2 = 4.982$

46. est. $\sigma = 2.232$

# CHAPTER 11

1. hypothesis

2. This is an unacceptable scientific research hypothesis because it is not falsifiable.

3. independent

4. dependent

5. subject variables

6. Yes, it is an experiment because there is an independent variable and a dependent variable.

7. The independent variable is the type of picture viewed--a picture of a clothed model or a nude model. This is the variable that is manipulated by the experimenter.

8. The dependent variable is the temperature of the subject's skin. This is the behavioral measure used by the experimenter.

9. The subject variable is gender, male or female. The subjects cannot be randomly assigned to gender because it is innate; thus, it is a subject variable.

10. No, this is not an experiment. There are no independent variables.

11. None. There are no independent variables; there is just the subject variable of gender.

12. The number of depressive episodes is the dependent variable because this is the behavioral measure used by the researcher.

13. The subject variable in this experiment is gender. Gender is a subject variable because this is an innate characteristic of a subject, and it cannot be assigned by the experimenter.

14. Yes, this is an experiment because it has an independent variable.

15. The independent variable is the type of pattern movement shown to the subject--vertical, horizontal, or rotational. It is an independent variable because the experimenter controls which type of pattern movement is seen by the subjects.

16. The dependent variable is the number of pattern elements remembered because this is the measure used by the experimenter.

17. None, there are no subject variables in this experiment because all the variables involved are either controlled by the experimenter or are behavioral measures. None of the variables are innate.

18. This is a completely randomized factorial design because there are two independent variables, each with at least two levels.

19. This is a completely randomized design because there is only one independent variable with two or more levels.

20. This is a one group design because the mean of a single sample is compared to the population mean.

21. This is a completely randomized design because there is one independent variable with two levels, the one being applied to the experimental group and the one being applied to the control group.

22. This is a completely randomized factorial design because there are at least two independent variables, each with two or more levels.

## CHAPTER 12

1. $$\overline{GPA} = \frac{31.1}{10} = 3.11$$

2. $S = .394$

3. est. $\sigma = .415$

4. $$\text{est. } \sigma_{\overline{X}} = \frac{.394}{\sqrt{10-1}} = .131$$

OR

$$\text{est. } \sigma_{\overline{X}} = \frac{.415}{\sqrt{10}} = .131$$

5. $$t = \frac{3.11 - 2.9}{.131} = 1.603$$

6. $df = 10 - 1 = 9$

7. Critical value (one-tailed) = 1.860

   No, it is not significant.

8. We fail to reject the null hypothesis that there is no difference between the GPA of returning students and the GPA of regular students.

9. $\bar{X}_1 = 61.8$
$\bar{X}_2 = 40.2$

10. $S_1 = 11.161$
$S_2 = 9.918$

11. est. $\sigma_{\bar{X}_1} = 3.721$
est. $\sigma_{\bar{X}_2} = 3.306$

12. est. $\sigma_{Diff} = 4.977$

13. $$t = \frac{61.8 - 40.2}{4.977} = 4.340$$

14. $df = 10 - 1 = 9$

15. Critical value = 2.262

Yes, the $t$ is significant.

16. Subjects eat a significantly larger amount of popcorn when they are watching a scary movie.

17 and 18.

| Child | Verbal | Nonverbal | $D$ | $D^2$ |
|---|---|---|---|---|
| C. J. | 3 | 6 | -3 | 9 |
| F. K. | 5 | 8 | -3 | 9 |
| M. O. | 7 | 9 | -2 | 4 |
| I. M. | 4 | 8 | -4 | 16 |
| G. G. | 2 | 4 | -2 | 4 |
| K. T. | 1 | 1 | 0 | 0 |
| B. W. | 4 | 3 | 1 | 1 |
| M. B. | 2 | 8 | -6 | 36 |

19. $\Sigma D = -19$

20. $\Sigma D = 79$

21. $\bar{D} = -2.375$

22.

$$\text{est. } \sigma_{\text{Diff}} = \sqrt{\frac{\frac{79}{8} - (-2.375)^2}{7}} = .778$$

23.

$$t = \frac{-2.375}{.788} = 3.053$$

24. $df = 7$

25. Critical value = 2.365

Yes, the $t$ is significant.

26. The nonverbal presentation of the math problems led to a significantly greater number of correctly solved problems.

27.

$$\sigma_{\overline{X}} = \frac{2}{\sqrt{16}} = \frac{2}{4} = .5$$

28.

$$t = \frac{5 - 7}{.5} = -4$$

29. $df = 15$

30. Critical value = 2.131

Yes, this is significant.

31. The new style hammer results in significantly fewer misses than the traditional hammer.

32. est. $\sigma_{\overline{X}_1} = 2$

est. $\sigma_{\overline{X}_2} = 3$

33. $\text{est. } \sigma_{\text{Diff}} = \sqrt{2^2 + 3^2 - (2 \cdot .8 \cdot 2 \cdot 3)}$

$= 1.844$

34. $t = \dfrac{47 - 52}{1.844} = -2.71$

35. $df = 9$

36. Critical value = 2.262

Yes, this is significant.

37. The self-hypnosis produced kicks of significantly greater distance.

## CHAPTER 13

Dosage condition

| A | | B | | C | |
|---|---|---|---|---|---|
| $X_1$ | $X_1^2$ | $X_2$ | $X_1^2$ | $X_3$ | $X_1^2$ |

$\Sigma X_1 = 105$ $\Sigma X_2 = 74$ $\Sigma X_3 = 33$

$\Sigma X_1^2 = 2263$ $\Sigma X_2^2 = 1110$ $\Sigma X_3^2 = 235$

$\Sigma\Sigma X = 212$ $\Sigma\Sigma X^2 = 3608$

1. $SS_{\text{bg}} = \dfrac{105^2}{5} + \dfrac{74^2}{5} + \dfrac{33^2}{5} - \dfrac{(212)^2}{15}$

$= 521.733$

2. $$SS_{total} = (2263 + 1110 + 3608) - \frac{(212)^2}{15}$$

$$= 611.733$$

3. $$SS_{wg} = 3608 - \left(\frac{105^2}{5} + \frac{74^2}{5} + \frac{33^2}{5}\right)$$

$$= 90$$

4. $df_{bg} = 3 - 1 = 2$

5. $df_{wg} = (5 - 1) + (5 - 1) + (5 - 1) = 12$

6. $df_{total} = 15 - 1 = 14$

7. $$MS_{bg} = \frac{521.733}{2} = 260.867$$

8. $$MS_{wg} = \frac{90}{12} = 7.5$$

9. $$F = \frac{260.867}{7.5} = 34.782$$

10. Critical value = 3.88

Yes, this is significant.

11. $$HSD = 3.77\sqrt{\frac{7.5}{5}} = 4.617$$

12. Given the significant $F$ and an $HSD$ of 4.617, we can conclude that there is a significant difference between all possible pairs of means. Dosage condition C resulted in significantly fewer numbers of delusions than conditions A or B, so if I were the psychiatrist, I would prescribe dosage level C.

Type of tape

| Audible $X_1$ | $X_1^2$ | Subliminal $X_2$ | $X_2^2$ | Control $X_3$ | $X_3^2$ |
|---|---|---|---|---|---|

$\Sigma X_1 = 282$ $\Sigma X_2 = 154$ $\Sigma X_3 = 146$

$\Sigma X_1^2 = 16016$ $\Sigma X_2^2 = 4758$ $\Sigma X_3^2 = 4290$

$\Sigma\Sigma X = 582$ $\Sigma\Sigma X^2 = 25064$

13. $SS_{bg} = 2329.6$
14. $SS_{total} = 2482.4$
15. $SS_{wg} = 152.8$
16. $df_{bg} = 2$
17. $df_{wg} = 12$
18. $df_{total} = 14$
19. $MS_{bg} = 1164.8$
20. $MS_{wg} = 12.733$
21. $F = 91.476$
22. Critical value = 3.88

    Yes, this is significant.

23. $HSD = 6.016$

24. It is clear from these data that the audible tape has a significantly greater effect on the memory task than the subliminal or control conditions. There was no significant difference between the subliminal and control conditions.

Type of shape

| Triangle | | Pentagon | | Octagon | | Dodecagon | |
|---|---|---|---|---|---|---|---|
| $X_1$ | $X_1^2$ | $X_2$ | $X_2^2$ | $X_3$ | $X_3^2$ | $X_4$ | $X_4^2$ |

$\Sigma X_1 = 9$ $\quad \Sigma X_2 = 19$ $\quad \Sigma X_3 = 25$ $\quad \Sigma X_4 = 35$

$X_1^2 = 15$ $\quad \Sigma X_2^2 = 63$ $\quad \Sigma X_3^2 = 111$ $\quad \Sigma X_4^2 = 207$

$\Sigma\Sigma X = 88$ $\quad \Sigma\Sigma X^2 = 396$

25. $SS_{bg} = 63.790$

26. $SS_{total} = 77.630$

27. $SS_{wg} = 13.830$

28. $df_{bg} = 3$

29. $df_{wg} = 20$

30. $df_{total} = 23$

31. $MS_{bg} = 21.263$

32. $MS_{wg} = 0.692$

33. $F = 30.750$

34. Critical value = 3.10

Yes, this is significant.

35. $HSD = 1.345$

36. Given the significant $F$ and the $HSD$, all the shapes require significantly different numbers of fixations, with the exception of the pentagon/dodecagon comparison which was not significant.

# CHAPTER 14

| | Diet | | |
|---|---|---|---|
| | Low protein | High protein | Row Total |
| Raised alone | $\Sigma X = 332$<br>$\Sigma X^2 = 22082$ | $\Sigma X = 182$<br>$\Sigma X^2 = 6710$ | 514 |
| Raised in colony | $\Sigma X = 215$<br>$\Sigma X^2 = 9267$ | $\Sigma X = 133$<br>$\Sigma X^2 = 3555$ | 348 |
| Column total | 547 | 315 | |

$\Sigma\Sigma X = 862$ $\Sigma\Sigma X^2 = 41614$

1.
$$SS_{total} = 22082 + 6710 + 9267 + 3555 - \frac{(332+182+215+113)^2}{20}$$

$$= 41614 - 37152.2 = 4461.8$$

2.

$$SS_{wg} = 41614 - \left[\frac{332^2}{5} + \frac{182^2}{5} + \frac{215^2}{5} + \frac{133^2}{5}\right]$$

$$= 41614 - 41452.4 = 161.6$$

3.

$$SS_r = \left[\frac{514^2}{10} + \frac{348^2}{10}\right] - 37152.2$$

$$= 38530 - 37152.2 = 1377.8$$

4.

$$SS_c = \left[\frac{547^2}{10} + \frac{315^2}{10}\right] - 37152.2$$

$$= 39843.4 - 37152.2 = 2691.2$$

5. $SS_{rXc} = 44618 - (161.6 + 1377.8 + 2691.2)$

$= 231.2$

6. $df_{total} = 20 - 1 = 19$

7. $df_{wg} = (5 - 1) + (5 - 1) + (5 - 1) + (5 - 1)$

$= 16$

8. $df_r = 2 - 1 = 1$

9. $df_c = 2 - 1 = 1$

10. $df_{rXc} = 1 \cdot 1 = 1$

11.

$$MS_{wg} = \frac{161.6}{16} = 10.10$$

12. $$MS_{r} = \frac{1377.8}{1} = 1377.8$$

13. $$MS_{c} = \frac{2691.2}{1} = 2691.2$$

14. $$MS_{rXc} = \frac{231.2}{1} = 231.2$$

15. $$F_{r} = \frac{1377.8}{10.10} = 136.416$$

16. $$F_{c} = \frac{2691.2}{10.10} = 266.455$$

17. $$F_{rXc} = \frac{231.2}{10.10} = 22.891$$

18. (1, 16)

    Critical value = 4.49

    Yes, the *F* ratio for rows is significant.

19. (1, 16)

    Critical value = 4.49

    Yes, the *F* ratio for columns is significant.

20. (1, 16)

    Critical value = 4.49

    Yes, the *F* ratio for the interaction is significant.

21.

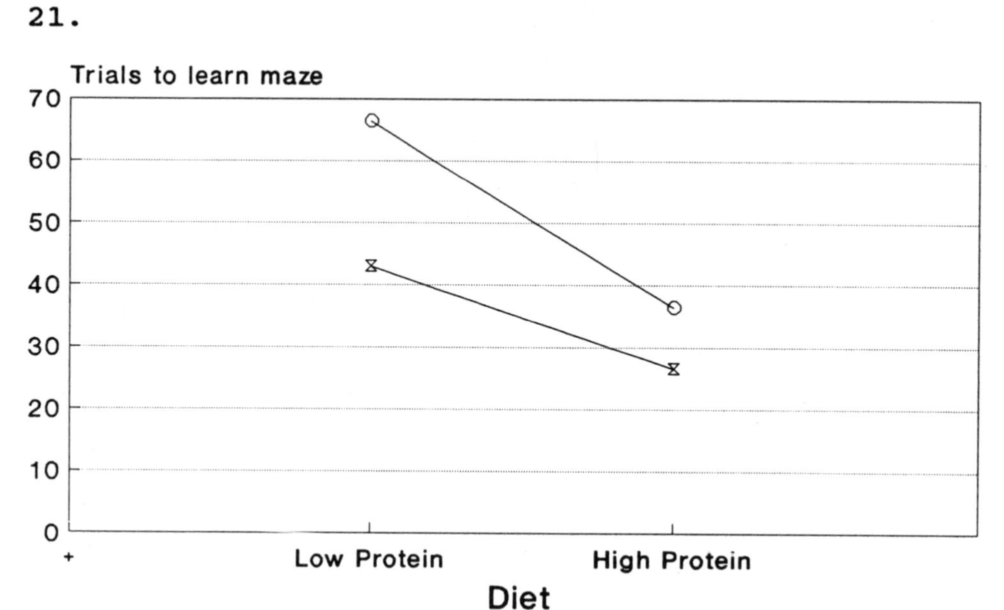


22. Both the diet and the environment in which the rats were raised had significant effects on the ability of the rats to run the maze. The rats took significantly fewer trials to run the maze when they were on a high protein diet. Rats raised in the colony also took significantly fewer trials to learn the maze. Looking at the graph of the interaction, it appears that the high protein diet had its biggest effect on those rats who were raised alone.

| | Loudness | | | |
|---|---|---|---|---|
| | Soft | Medium | Loud | Row Total |
| Same ear | $\Sigma X = 35$<br>$\Sigma X^2 = 409$ | $\Sigma X = 51$<br>$\Sigma X^2 = 869$ | $\Sigma X = 63$<br>$\Sigma X^2 = 1325$ | 149 |
| Diff. ears | $\Sigma X = 22$<br>$\Sigma X^2 = 166$ | $\Sigma X = 42$<br>$\Sigma X^2 = 590$ | $\Sigma X = 53$<br>$\Sigma X^2 = 941$ | 117 |
| Column total | 57 | 93 | 116 | |

$\Sigma\Sigma X = 266$ $\Sigma\Sigma X^2 = 4300$

23. $SS_{total} = 369.111$

24. $SS_{wg} = 16$

25. $SS_{r} = 56.889$

26. $SS_{c} = 294.778$

27. $SS_{rXc} = 1.445$

28. $df_{total} = 17$

29. $df_{wg} = 12$

30. $df_{r} = 1$

31. $df_{c} = 2$

32. $df_{rXc} = 2$

33. $MS_{wg} = 1.333$

34. $MS_{r} = 56.889$

35. $MS_{c} = 147.389$

36. $MS_{rXc} = 0.722$

37. $F_{r} = 42.667$

38. $F_{c} = 110.542$

39. $F_{rXc} = 0.542$

40. (1, 12)

    Critical value = 4.75

    Yes, this is significant.

41. (2, 12)

    Critical value = 3.88

    Yes, this is significant.

42. *HSD* = 1.777

    All column means are significantly different from one another.

43. (2, 12)

    Critical value = 3.88

    No, this is not significant.

44. There is no graph since the interaction is not significant.

45. Both of the main effects are statistically significant. The subjects were more accurate at determining the relative pitch when the tones were presented to the same ear rather than to opposite ears. Subjects were more accurate at determining the relative pitch when the tone was higher rather than lower.

# CHAPTER 15

1-4. The answers to problems 1 through 4 can be found in the following table.

| | Should psychologists prescribe drugs? Yes | No | Total |
|---|---|---|---|
| Psychologists | A $f_O=125$ $f_e=87.5$ | B $f_O=25$ $f_e=62.5$ | 150 |
| Medical doctors | C $f_O=50$ $f_e=87.5$ | D $f_O=100$ $f_e=62.5$ | 150 |
| Total | 175 | 125 | Grand total=300 |

5.

$$X^2 = \frac{(125-87.5)^2}{87.5} + \frac{(25-62.5)^2}{62.5} + \frac{(50-87.5)^2}{87.5} + \frac{(100-62.5)^2}{62.5}$$

$$= 16.071 + 22.5 + 16.071 + 22.5 = 77.142$$

6. $df = (2 - 1) \cdot (2 - 1) = 1$

7. Critical value = 3.841

   Yes, this is significant.

8. Medical doctors and psychologists have significantly different opinions about granting psychologists the right to prescribe psychoactive drugs.

9-12. The answers to problems 9 through 12 can be found in the following table.

| | Believe in ESP? | | | |
|---|---|---|---|---|
| Grad Students | Yes | No | Maybe | Total |
| Psychology | $f_o$= 11<br>$f_e$=25.97 | $f_o$= 25<br>$f_e$=8.59 | $f_o$= 5<br>$f_e$=6.44 | 41 |
| Sciences | $f_o$= 50<br>$f_e$=44.35 | $f_o$= 10<br>$f_e$=14.66 | $f_o$= 10<br>$f_e$=10.99 | 70 |
| Humanities | $f_o$= 60<br>$f_e$=50.68 | $f_o$= 5<br>$f_e$=16.75 | $f_o$= 15<br>$f_e$=12.57 | 80 |
| Total | 121 | 40 | 30 | Grand Total=191 |

13. $X^2$ = 53.017

14. $df$ = 4

15. Critical value = 9.488

Yes, it is significant.

16.

| Scary $X_1$ | $R_1$ | Musical $X_2$ | $R_2$ |
|---|---|---|---|
| 45 | 7 | 32 | 2 |
| 67 | 16 | 38 | 5 |
| 69 | 17 | 33 | 3 |
| 56 | 12 | 49 | 9.5 |
| 73 | 18 | 44 | 6 |
| 56 | 12 | 60 | 14 |
| 63 | 15 | 48 | 8 |
| 84 | 19 | 36 | 4 |
| 49 | 9.5 | 23 | 1 |
| 56 | 12 | | |
| | $\Sigma R_1 = 137.5$ | | $\Sigma R_2 = 52.5$ |

17. $\Sigma R_1 = 137.5$

18. $\Sigma R_2 = 52.5$

19. $$U_1 = (10 \cdot 9) + \frac{10 \cdot (10 + 1)}{2} - 137.5$$

$$= 7.5$$

20. $$U_2 = (9 \cdot 10) + \frac{9 \cdot (9 + 1)}{2} - 52.5$$

$$= 82.5$$

21. $U = 7.5$

22. Critical value = 20

Yes, the computed $U$ is significant.

23. Subjects eat more popcorn at scary movies than at musicals.

24.

| New | $R_1$ | Traditional | $R_2$ |
|---|---|---|---|
| 84 | 6.5 | 88 | 11 |
| 80 | 1 | 90 | 13.5 |
| 88 | 11 | 85 | 8 |
| 91 | 15.5 | 88 | 11 |
| 90 | 13.5 | 84 | 6.5 |
| 83 | 4 | 87 | 9 |
| 83 | 4 | 92 | 17 |
| | | 91 | 15.5 |
| | | 83 | 4 |
| | | 82 | 2 |
| | $\Sigma R_1 = 55.5$ | | $\Sigma R_2 = 97.5$ |

25. $\Sigma R_1 = 55.5$

26. $\Sigma R_2 = 97.5$

27. $U_1 = 42.5$

28. $U_2 = 27.5$

29. $U = 27.5$

30. Critical value = 14

    No, this is not significant.

31. Since the $U$ is not significant, there is no reason to believe that there is any significant difference between the two types of hammers.

32-35. The answers to problems 32 through 35 are found in the following table.

| Child | Verbal | Nonverbal | *D* | +R | -R |
|---|---|---|---|---|---|
| C. J. | 3 | 6 | -3 | | 4.5 |
| F. K. | 5 | 8 | -3 | | 4.5 |
| M. O. | 7 | 9 | -2 | | 2.5 |
| I. M. | 4 | 8 | -4 | | 6 |
| G. G. | 2 | 4 | -2 | | 2.5 |
| ~~K. T.~~ | ~~1~~ | ~~1~~ | | | |
| B. W. | 4 | 3 | 1 | 1 | |
| M. B. | 2 | 8 | -6 | | 7 |
| | | | | Σ+R= 1 | Σ-R= 27 |

36. $T = 1$

37. $n = 7$

38. Critical value = 2

Yes, this is significant.

39-42. The answers to problems 39 through 42 are found in the following table.

| Punter | Normal | Hypno | *D* | +R | -R |
|---|---|---|---|---|---|
| C. J. | 45 | 48 | - 3 | | 3 |
| F. K. | 40 | 45 | - 5 | | 5 |
| M. O. | 33 | 32 | - 1 | | 1 |
| I. M. | 35 | 39 | - 4 | | 4 |
| G. G. | 47 | 53 | - 6 | | 6.5 |
| K. T. | 52 | 65 | -13 | | 8 |
| B. W. | 47 | 53 | 6 | 6.5 | |
| ~~M. B.~~ | ~~46~~ | ~~46~~ | ~~0~~ | | |
| K. T. | 39 | 41 | - 2 | | 2 |
| B. B. | 28 | 45 | -17 | | 9 |
| | | | | Σ+R= 6.5 | Σ-R=38.5 |

43. $T = 6.5$

44. $n = 9$

45. Critical value = 5

No, this is not significant.

46. Since the computed value of $T$ is greater than the critical value of 5, we must fail to reject the null hypothesis and conclude that there is no significant difference between the distances the punters kick the football while under the the two different levels of consciousness.

47-48. The answers to problems 47 and 48 are shown in the following table.

Type of Tape

| Audible | | Subliminal | | Control | |
|---|---|---|---|---|---|
| $X_1$ | $R_1$ | $X_2$ | $R_2$ | $X_3$ | $R_3$ |
| 55 | 17 | 32 | 9.5 | 30 | 6.5 |
| 62 | 21 | 30 | 6.5 | 29 | 4.5 |
| 49 | 15 | 28 | 2.5 | 26 | 1 |
| 55 | 17 | 31 | 8 | 33 | 12 |
| 61 | 20 | 33 | 12 | 28 | 2.5 |
| 58 | 19 | 32 | 9.5 | 29 | 4.5 |
| 55 | 17 | 36 | 14 | 33 | 12 |
| $\Sigma R_1 =$ | 126 | $\Sigma R_2 =$ | 62 | $\Sigma R_3 =$ | 43 |

49. $n_1 = 7$

$n_2 = 7$

$n_3 = 7$

$N_T = 21$

50.

$$H = \left[\frac{12}{21\cdot(21+1)}\right]\cdot\left[\frac{126^2}{7} + \frac{62^2}{7} + \frac{43^2}{7}\right] - 3\cdot(21 + 1)$$

$= 14.113$

51. $df = 3 - 1 = 2$

Critical value = 5.991

Yes, this is significant.

52. There was a significant difference among the three different conditions on the memory test.

53-54. The answers for problems 53 and 54 can be found in the following table.

Shape of object

| Triangle | | Pentagon | | Octagon | | Dodecagon | |
|---|---|---|---|---|---|---|---|
| $X_1$ | $R_1$ | $X_2$ | $R_2$ | $X_3$ | $R_3$ | $X_4$ | $R_4$ |
| 1 | 2 | 2 | 6 | 2 | 6 | 5 | 18 |
| 2 | 6 | 3 | 10 | 4 | 13.5 | 6 | 22 |
| 1 | 2 | 4 | 13.5 | 5 | 18 | 6 | 22 |
| 2 | 6 | 4 | 13.5 | 5 | 18 | 6 | 22 |
| 2 | 6 | 3 | 10 | 4 | 13.5 | 5 | 18 |
| 1 | 2 | 3 | 10 | 5 | 18 | 7 | 24 |
| $\Sigma R_1 = 24$ | | $\Sigma R_2 = 63$ | | $\Sigma R_3 = 87$ | | $\Sigma R_4 = 126$ | |

55. $n_1 = 6$

$n_2 = 6$

$n_3 = 6$

$n_4 = 6$

$N_T = 24$

56. $H = 18.3$

57. $df = 3$

Critical value = 7.815

Yes, this is significant.

58. The results indicate that there are significant differences in the number of eye movements required to identify the four shapes.